AF333440

Reevaluation of the State of the Science for Water-Quality Criteria Development

Other titles from the Society of Environmental Toxicology and Chemistry (SETAC):

Porewater Toxicity Testing: Biological, Chemical, and Ecological Considerations
Carr and Nipper, editors
2003

Contaminated Soils: From Soil–Chemical Interactions to Ecosystem Management
Lanno, editor
2003

Silver in the Environment
Andren and Bober, editors
2002

Test Methods to Determine Hazards for Sparingly Soluble Metal Compounds in Soils
Fairbrother, Glazebrook, van Straalen, Tararzona, editors
2002

Bioavailability of Metals in Terrestrial Ecosystems: Importance of Partitioning for Bioavailability to Invertebrates, Microbes, and Plants
Allen, editor
2002

Interconnections between Human Health and Ecological Variability
Di Giulio and Benson, editors
2002

Avian Effects Assessment: A Framework for Contaminants Studies
Hart, Balluff, Barfknecht, Chapman, Hawkes, Joermann, Leopold, Luttik, editors
2001

Ecological Variability: Separating Natural from Anthropogenic Causes of Ecosystem Impairment
Baird and Burton, editors
2001

Guidance Document on Regulatory Testing and Risk Assessment Procedures for Protection Products with Non-Target Arthropods
Candolfi, Barrett, Campbell, Forster, Grady, Huet, Lewis, Schmuck, Vogt, editors
2001

Test Methods to Determine Hazards of Sparingly Soluble Metals in Soils
Fairbrother, Glazebrook, van Straalen, Tarazona, editors
2002

Risk Management: Ecological Risk-Based Decision-Making
Stahl, Bachman, Barton, Clark, deFur, Ells, Pittinger, Slimak, Wentsel, editors
2001

Ecological Risk Assessment of Contaminated Sediments
Ingersoll, Dillon, Biddinger, editors
1997

For information about any SETAC publication, including SETAC's international journal, *Environmental Toxicology and Chemistry*, contact the SETAC Office nearest you:
Administrative Offices

SETAC North America
1010 North 12th Avenue
Pensacola, Florida 32501-3367 USA
T 850 469 1500; F 850 469 9778
E setac@setac.org

SETAC Europe
Avenue de la Toison d'Or 67
B-1060 Brussels, Belgium
T 32 2 772 72 81; F 32 2 770 53 86
E setac@setaceu.org

www.setac.org

Environmental Quality Through Science®

Reevaluation of the State of the Science for Water-Quality Criteria Development

Edited by

Mary C. Reiley
U.S. Environmental Protection Agency
Washington, DC, USA

William A. Stubblefield
Parametrix, Inc.
Corvallis, Oregon, USA

William J. Adams
Rio Tinto
Murray, Utah, USA

Dominic M. Di Toro
University of Delaware
Newark, Delaware, USA

Peter V. Hodson
Queen's University
Kingston, Ontario, Canada

Russell J. Erickson
U.S. Environmental Protection Agency
Duluth, Minnesota, USA

F. James Keating Jr.
U.S. Environmental Protection Agency
Washington, DC, USA

Coordinating Editor of SETAC Books
Andrew Green
International Lead Zinc Research Organization
Department of Environment and Health
Research Triangle Park, North Carolina, USA

Published by the Society of Environmental Toxicology and Chemistry (SETAC)

Cover design by Michael Kenney Graphic Design and Advertising
Index by Celia McCoy

Library of Congress Cataloging-in-Publication Data

Pellston Workshop on Reevaluation of the State of the Science for Water-Quality Criteria Development (1998 : Fairmont Hot Springs, Mont.)
　　　Reevaluation of the state of the science for water-quality criteria development : proceedings from the Pellston Workshop on Reevaluation of the State of the Science for Water-Quality Criteria Development, 25-30 June 1998, Fairmont Hot Springs, Montana / edited by Mary Reiley ...[et al.]
　　　p. cm. -- (SETAC special publications series)
　　　"Publication sponsored by the Society of Environmental Toxicology and Chemistry (SETAC) and the SETAC Foundation for Environmental Education"
Includes bibliographic references and index.
ISBN 1-880611-30-9 (alk. paper)
　　　1. Water quality bioassay--Congresses. 2. Ecological risk assessment--Congresses.　　I. Reiley, Mary, 1964- II. SETAC (Society) III. SETAC Foundation for Environmental Education.　　IV. Title.　　V. Series.

QH90.57.B5 P45 2003
363.739'46--dc21　　　　　　　　　　　　　　　　　　　　　　　　　　　　　　00-020854

Information in this book was obtained from individual experts and highly regarded sources. It is the publisher's intent to print accurate and reliable information, and numerous references are cited; however, the authors, editors, and publisher cannot be responsible for the validity of all information presented here or for the consequences of its use. Information contained herein does not necessarily reflect the policy or views of the Society of Environmental Toxicology and Chemistry (SETAC). Mention of commercial or noncommercial products and services does not imply endorsement or affiliation by the author or SETAC.

No part of this publication may be reproduced, stored in a retrieval system, or transmitted in any form or by any means, electronic, electrostatic, magnetic tape, mechanical, photocopying, recording, or otherwise, without permission in writing from the copyright holder.

All rights reserved. Authorization to photocopy items for internal or personal use, or for the personal or internal use of specific clients, may be granted by the Society of Environmental Toxicology and Chemistry (SETAC), provided that the appropriate fee is paid directly to the Copyright Clearance Center, Inc., 222 Rosewood Drive, Danvers, MA 01923 USA (Telephone 978 750 8400) or to SETAC. Before photocopying items for educational classroom use, please contact the Copyright Clearance Center (http://www.copyright.com) or the SETAC Office in North America (Telephone 850 469 1500, Fax 850 469 9778, E-mail setac@setac.org). SETAC's consent does not extend to copying for general distribution, for promotion, for creating new works, or for resale. Specific permission must be obtained in writing from SETAC for such copying. Direct inquiries to the Society of Environmental Toxicology and Chemistry (SETAC), 1010 North 12th Avenue, Pensacola, FL 32501-3367, USA.

© 2003 Society of Environmental Toxicology and Chemistry (SETAC)
This publication was printed on recycled paper using soy ink.
SETAC Press is an imprint of the Society of Environmental Toxicology and Chemistry.
No claim is made to original U.S. Government works.

International Standard Book Number 1-880611-30-9
Printed in the United States of America
10 09 08 07 06 05 04 03　　　10 9 8 7 6 5 4 3 2 1

∞ The paper used in this publication meets the minimum requirements of the American National Standard for Information Sciences — Permanence of Paper for Printed Library Materials, ANSI Z39.48-1984.

Reference Listing: Reiley MC, Stubblefield WA, Adams WJ, Di Toro DM, Hodson PV, Erickson RJ, Keating Jr FJ, editors. 2003. Reevaluation of the State of the Science for Water-Quality Criteria Development. Pensacola, FL, USA: Society of Environmental Toxicology and Chemistry (SETAC). 224 p.

SETAC Publications

Books published by the Society of Environmental Toxicology and Chemistry (SETAC) provide in-depth reviews and critical appraisals on scientific subjects relevant to understanding the impacts of chemicals and technology on the environment. The books explore topics reviewed and recommended by the SETAC Publications Advisory Council and approved by the appropriate governing body for their importance, timeliness, and contribution to multidisciplinary approaches to solving environmental problems. The diversity and breadth of subjects covered in the publications reflect the wide range of disciplines encompassed by environmental toxicology, environmental chemistry, and hazard and risk assessment, and life-cycle assessment. SETAC books attempt to present the reader with authoritative coverage of the literature, as well as paradigms, methodologies, and controversies; research needs; and new developments specific to the featured topics. The books are generally peer reviewed for SETAC by acknowledged experts.

SETAC publications, which include Technical Issue Papers (TIPs), workshop summaries, newsletter (*SETAC Globe*), and journal (*Environmental Toxicology and Chemistry*), are useful to environmental scientists in research, research management, chemical manufacturing and regulation, risk assessment, life-cycle assessment, and education, as well as to students considering or preparing for careers in these areas. The publications provide information for keeping abreast of recent developments in familiar subject areas and for rapid introduction to principles and approaches in new subject areas.

SETAC recognizes and thanks the past SETAC books editors:

> C.G. Ingersoll, Midwest Science Center
> > U.S. Geological Survey, Columbia, MO, USA

> T.W. La Point, Institute of Applied Sciences
> > University of North Texas, Denton, TX, USA

> B.T. Walton, U.S. Environmental Protection Agency
> > Research Triangle Park, NC, USA

> C.H. Ward, Department of Environmental Sciences and Engineering
> > Rice University, Houston, TX, USA

Contents

CHAPTER 1

Problem Formulation 1

Thomas W. La Point, Scott E. Belanger, Trudie Crommentuijn,
John Goodrich-Mahoney, Robert A. Kent, Donald I. Mount, Douglas J. Spry,
Torgny Vigerstad, Dominic M. Di Toro, F. James Keating Jr, Mary C. Reiley

CHAPTER 2

Exposure Analysis 15

William H. Benson, Herbert E. Allen, John P. Connolly, Charles G. Delos,
Lenwood W. Hall Jr, Samuel N. Luoma, David Maschwitz, Joseph S. Meyer,
John W. Nichols, William A. Stubblefield

List of Figures

List of Tables

Acknowledgments

This workshop would not have been possible without the support of the Society of Environmental Toxicology and Chemistry (SETAC), the SETAC Foundation for Environmental Education, and corporate sponsors. Linda Longsworth, Greg Schiefer, and Leslie Long are greatly appreciated for their efforts in planning and conducting the workshop. Thanks also to the support of the SETAC Office in Pensacola for the production and publication of this book. We wish to acknowledge the peer reviewers, Peter M. Chapman and Richard A. Kimerle, whose insight and comments helped make the document as well-rounded and balanced as it is.

Corporate Sponsors

Atlantic Richfield Company

Eastman Kodak

Electric Power Research Institute

Environment Canada

International Copper Association

Kennecott Utah Copper Corporation

U.S. Environmental Protection Agency, Office of Water and Office of Science

About the Editors

Mary C. Reiley, MS, is the Program Manager for the National Ambient Water Quality Criteria Program in the U.S. Environmental Protection Agency (USEPA). Reiley guides two teams that develop national ambient water-quality criteria for aquatic life and human health, provide technical assistance for criteria derivation and implementation to states, tribes, and other stakeholders, and conduct investigations into the fate, transport, bioavailability, and effects of toxics to aquatic life and human health. She also coordinates research agendas between the Office of Water and the Office of Research and Development and is developing Agency-wide factors for assessing the quality of third party information. Previously she has directed the contaminated sediment program and developed national water-quality enforcement policy. Mary holds an MS in Environmental Biology from George Mason University and a BS in Science-Biology from the College of William and Mary.

William A. Stubblefield, PhD, is a senior environmental toxicologist with Parametrix, Inc. Toxicology group in Corvallis, Oregon, USA. He also serves on the faculty of Oregon State University's Department of Molecular and Environmental Toxicology. Dr. Stubblefield is an active member of the Society of Environmental Toxicology and Chemistry (SETAC), where he currently serves as chairman of the Publications Advisory Council, associate editor of the Society's newsletter, and Vice President of the Society's Board of Directors.

Dr. Stubblefield received a PhD in Aquatic Toxicology from the University of Wyoming, 1987; an MS in Toxicology/Toxicodynamics from the University of Kentucky, 1979; and a BS in Biological Sciences/Chemistry from Eastern Kentucky University, 1977.

Dr. Stubblefield has 20-plus years experience in environmental toxicology, environmental assessment, and aquatic and wildlife toxicology studies. He has dealt with a number of related environmental issues, especially those associated with the evaluation of impacts to aquatic and terrestrial species resulting from the discharge of mine-associated waters and tailings, and the toxicity of metals and byproducts of metal mining.

William J. Adams, PhD, is currently Principal Environmental Scientist for Rio Tinto, Salt lake City, Utah, USA. He was previously Director of Environmental Science for six years at Kennecott Utah Copper, Vice President of ABC Laboratories for 5 years and Science Fellow at Monsanto Company for 14 years. Recent research interests include developing ecotoxicology risk assessment methods for metals, site-specific methodologies for water-quality criteria for metals, and development of a persistent, bioaccumulative, and toxic (PBT) approach for hazard assessment of metals. Dr. Adams has published several papers on methods for assessing sediments and was instrumental in developing the science supporting equilibrium partitioning theory for nonpolar organic substances. He has also published several papers in the area of water-quality assessments. Dr. Adams is a member of the U.S. Environmental Protection Agency (USEPA) Science Advisory Board and the USEPA Superfund National Advisory Committee for Environmental Policy and Technology.

Dominic M. Di Toro, PhD, is presently Distinguished Professor of Civil and Environmental Engineering at the University of Delaware. Previously, he was the Donald J. O'Connor Professor of Environmental Engineering at Manhattan College. He is also a Principal Consultant at HydroQual, Inc. Dr. Di Toro has a BEE in Electrical Engineering from Manhattan College, and an MA in Electrical Engineering and a PhD in Civil and Geological Engineering from Princeton University. Dr. Di Toro has specialized in the development and application of mathematical and statistical analyses to stream, lake, estuarine, and coastal water-quality and sediment problems. Recently, his work has focused on the development of metals and toxic organic chemical water- and sediment-quality criteria for the USEPA, sediment flux models for nutrients and metals, and integrated hydrodynamic, sediment transport, and water quality models. He has appeared before the USEPA Science Advisory Board, together with USEPA and other academic scientists, on 3 occasions to present the Equilibrium Partitioning methodology for developing sediment criteria for metals and organics, and most recently to present the Biotic Ligand Model for the development of water-quality criteria for metals.

Peter V. Hodson, PhD, is the Director of the School of Environmental Studies and a Professor of Biology at Queen's University, Canada. He has a BSc from McGill University (1968), an MSc from the University of New Brunswick (1970), and a PhD from the University of Guelph (1974). From 1974 to 1995, he was a scientist with Fisheries and Oceans and with Environment Canada, and he joined Queen's in 1995. Dr. Hodson has authored more than 100 technical publications in fish toxicology. His research has contributed to programs of water-quality management, most notably water-quality objectives for the Great Lakes, pulp mill effluent regulations, environmental effects monitoring programs, and policies on chlorine. He is currently studying the ecological risks of sediments contaminated by oil and coal tar. Dr. Hodson is a Past President of SETAC and has served on its Board of Directors, as program Chair of SETAC's 10th Annual Meeting in Toronto, 1989, and as an editor of *Environmental Toxicology and Chemistry*. He has contributed to many technical panels, particularly those of the International Joint Commission, American Fisheries Society, Department of Fisheries and Oceans, and Environment Canada.

Russell J. Erickson, PhD, is a research chemist at the Mid-Continent Ecology Division of the National Health and Environmental Effects Research Laboratory of the USEPA. Dr. Erickson received BS degrees in fisheries and wildlife and in botany from Michigan State University in 1973; a MS degree in fisheries and wildlife/limnology from Michigan State University in 1976; and a PhD degree in oceanography and limnology/water chemistry from the University of Wisconsin-Madison in 1981. His research has addressed the influence of exposure conditions on toxicity of chemicals to aquatic organisms, including effects of physicochemical variables on metal and ammonia toxicity; regulation of uptake of organic chemicals at fish gills; relationships among toxicity, chemical accumulation, and exposure time-series; metals toxicity via dietary exposure; and risks from photoactivated toxicity. He has assisted in various efforts by USEPA and other agencies regarding the development of water quality criteria and risk assessments for aquatic life.

F. James Keating Jr., MS, is an Environmental Scientist with the USEPA in Washington DC. He currently works on the development and implementation of policy, regulations, and guidance for the review and approval of state and tribal water-quality standards under the Clean Water Act. He has previously directed national assessments of contaminated sediment and mercury in fish tissue. His areas of expertise include human health and aquatic life risk assessment and the application of water-quality criteria for the protection of their uses. He holds an MS in Environmental Management from Duke University and a BSc from the University of Richmond, Virginia. Prior to joining the USEPA in 1995, he worked at Versar, Inc., a consulting and engineering firm in Springfield, Virginia, USA.

Workshop Participants†

William J. Adams*
Kennecott Utah Copper Corporation
Magna, UT, USA

Herbert E. Allen
University of Delaware
Newark, DE, USA

Gerald T. Ankley
U.S. Environmental Protection Agency
Duluth, MN, USA

Pierre Beland
IJC/Institut Nationale d'Ecotoxocologia
du St. Laurent
Ottawa, ON, Canada

Scott E. Belanger
The Procter and Gamble Company
Cincinnati, OH, USA

William H. Benson ‡
University of Mississippi
University, MS, USA

Kevin V. Brix
Parametrix, Inc.
Kirkland, WA, USA

William H. Clements
Colorado State University
Fort Collins, CO, USA

John P. Connolly
Quantitative Environmental Analysis, LLC
Montvale, NJ, USA

Trudie Crommentuijn
RIVM - Centre for Substances and Risk
Assessment
Bilthoven, Netherlands

Charles G. Delos
U.S. Environmental Protection Agency
Washington, DC, USA

Dominic M. Di Toro
Manhattan College
HydroQual, Inc.
Mahwah, NJ, USA

D. George Dixon
University of Waterloo
Ontario, Canada

Russell J. Erickson*
U.S. Environmental Protection Agency
Duluth, MN, USA

Anne Fairbrother
ep&t
Corvallis, OR, USA

Jeffrey M. Giddings
Springborn Labs
Wareham, MA, USA

John Goodrich-Mahoney
Electric Power Research Institute
Washington, DC, USA

Lenwood W. Hall
University of Maryland
Queenstown, MD, USA

David J. Hansen
HydroQual, Inc.
Mahwah, NJ, USA

Christopher W. Hickey
National Institute of Water and
Atmospheric Research
Hamilton, New Zealand

Peter V. Hodson*
Queen's University
Kingston, ON, Canada

F. James Keating Jr*
U.S. Environmental Protection Agency
Washington, DC, USA

Robert A. Kent
Environment Canada
Hull, PQ, Canada

Wayne G. Landis
Western Washington University
Bellingham, WA, USA

Roman P. Lanno
Oklahoma State University
Stillwater, OK, USA

Thomas W. La Point‡
Texas Tech University
Lubbock, TX, USA

Christopher M. Lee
International Copper Association
New York, NY, USA

Samuel N. Luoma
U.S. Geological Survey
Menlow Park, CA, USA

Beth L. McGee
U.S. Fish and Wildlife Service
Annapolis, MD, USA

Eugene R. Mancini
ARCO
Los Angeles, CA, USA

David Maschwitz
Minnesota Pollution Control Agency
St. Paul, MN, USA

Joseph S. Meyer
University of Wyoming
Laramie, WY, USA

Dwayne R.J. Moore‡
The Cadmus Group
Ottawa, ON, Canada

Donald I. Mount
AScI Corporation
Duluth, MN, USA

David R. Mount‡
U.S. Environmental Protection Agency
Duluth, MN, USA

Wayne R. Munns
U.S. Environmental Protection Agency
Narragansett, RI, USA

John W. Nichols
U.S. Environmental Protection Agency
Duluth, MN, USA

Mary C. Reiley*
U.S. Environmental Protection Agency
Washington, DC, USA

Robert K. Ringer
Michigan Sate University
Traverse City, MI, USA

Greg Schiefer*
SETAC/SETAC Foundation
Pensacola, FL, USA

Keith R. Solomon
Canadian Centre for Toxicology
Guelph, ON, Canada

Douglas J. Spry
Ontario Ministry of Environment
Toronto, ON, Canada

Jane P. Staveley
The Cadmus Group
Durham, NC, USA

William A. Stubblefield*
ENSR Consulting and Engineering
Fort Collins, CO, USA

John E. Toll
Parametrix, Inc.
Kirkland, WA, USA

Torgny Vigerstad
Bio-Response Systems
Dartmouth, NS, Canada

Christopher M. Wood
McMaster University
Hamilton, ON, Canada

William Wuerthele
U.S. Environmental Protection Agency
Denver, CO, USA

† Affiliations were current at the time of the workshop

* Steering Committee

‡ Chapter Chair

Foreword

The Organization of the Workshop and the Remaining Chapters

F. James Keating Jr.

Purpose

The purpose of this workshop was to evaluate how ecological risk assessment (ERA) concepts can be used to develop water-quality criteria (WQC) and water-quality methods in a more integrated and efficient process. Specifically, the risk assessment paradigm was used as an organization framework—comprising problem formulation, exposure assessment, effects assessment, and risk characterization. Participants were encouraged to think broadly and holistically about sound conceptual approaches. The primary question was, if one were to design a WQC approach from scratch, with the knowledge and data currently available or reasonably available in the near future, how would one do it? Questions that were asked included the following: What resources should be protected? What are the appropriate assessment and measurement endpoints? What level of protection should be provided? Other issues that were considered included the use of tissue residue effects data, dietary toxicity, sediment interaction, wildlife concerns, species sensitivity, missing data, role of biomarkers, probabilistic analyses, and population-level and community-level effects. The primary focus was on ecological receptors. Although exposure from consumption of aquatic organisms (for wildlife and humans) was considered, this workshop did not examine the human toxicological issues.

Workshop Groups

The groups into which the attendees were organized were based on the risk assessment paradigm. This is mirrored in the structure of the book. In addition, unlike the usual Pellston Workshop organization, 2 sets of workgroups were employed.

Phase 1 workgroups

In the first phase of the workshop, participants were distributed among four workgroups that worked independently on the same task: developing a conceptual model for the formulation and implementation of WQC. Although experiences with current regulatory procedures will help this process, these groups were asked to start with a "clean slate," to not be obligated to follow any of the specifics of current practice, and to "push the envelope" in terms of what criteria should be. This conceptual model was to address risks to water-column and sediment-dwelling aquatic organisms and risks to semiaquatic or terrestrial animals that feed on aquatic organisms. Effects on humans were not to be addressed, although procedures for assessing exposures to fish-consuming wildlife would be applicable to human risk assessments.

Such a conceptual model is the goal of the problem formulation stage of a risk assessment, and the groups were to first identify assessment endpoints that would be of concern to WQC and then attempt to specify a general framework of data, procedures, and models that could produce a measure of risk relevant to the endpoints. Groups were not to try to address a specific level of protection that criteria should achieve, but rather accommodate flexibility in the degree and nature of the risk. Issues that were to be addressed included

- the aquatic ecosystem to which criteria are to be applied, especially spatial extent, and the system attributes that criteria should address;
- the types of information that an exposure analysis should provide (i.e., the distribution of chemicals in space, time, and form) and the general data and tools needed in an exposure analysis;
- the types of information that an effects analysis should provide and the general data and tools needed in an effects analysis; and
- the measure of risk that should come from integrating exposure and effects information and the relationship of those measures to system attributes of interest.

Phase 2 workgroups

After the conceptual model development phase, the attendees were divided into four workgroups to elaborate on the problems, possible solutions, and data and research needs.

Problem formulation workgroup

This group carried on the efforts of the workgroups from Phase 1, refining and describing the conceptual model.

Exposure analysis workgroup

This group addressed procedures to describe and predict the exposure concentrations needed to assess effects. Points that were addressed included spatial and temporal variations in concentrations, chemical partitioning, and speciation that can affect bioavailability, other environmental factors that might affect toxicity, and bioaccumulation, which will dictate doses via food and exposures on an internal receptor basis.

Effects analysis workgroup

This group addressed procedures to describe responses of aquatic populations to chemical exposures. Points addressed included toxicity test data requirements, implications of temporal variability of exposure and organism susceptibility, effects of speciation and other physicochemical exposure conditions, relationship of toxicity to chemical accumulation, and multiple routes of exposure. This workgroup addressed the implications of chemical toxicity to population response, including the impact of temporal and spatial variations of toxicity.

Risk characterization workgroup

This group addressed procedures to integrate exposure and effects analysis into a useful measure of risk that can be related to assessment endpoints. Points addressed included the probabilistic combination of exposure and effects information, uncertainty analysis, and strategies to provide an integrated index of risk for biological populations and communities. This workgroup also addressed the relationship of risks based on toxicological data and models to observed responses of aquatic communities (i.e., validation).

Summary

The results of each group's deliberations are presented in the following chapters. They are the record of the work accomplished and the basis for the conclusions and recommendations presented at the end of each chapter.

Executive Summary

Dominic M. Di Toro

Water-quality criteria (WQC) were first developed in the 1950s and have emerged as one of the primary tools for managing surfacewater quality. Since then, there have been rapid developments in the science of aquatic toxicology and chemistry and the methods to derive a number of WQC to protect a variety of organisms and endpoints. There is a need to integrate and harmonize approaches for developing criteria for water, sediments, and tissues. In June 1998, a Society of Environmental Toxicology and Chemistry (SETAC) Pellston Workshop was held in Gregson, Montana, USA, to determine whether the current approaches for establishing and implementing WQC can be improved and whether the development of different types of criteria can be harmonized. Fifty experts from North America, Europe, and New Zealand identified problems with the current approach, assessed recent advances in environmental toxicology and chemistry, and recommended a 3-level system for broadly applicable and site-specific criteria.

Problems Exist with Current Criteria

Aquatic life criteria have been, and continue to be, among the most important tools available to assess and manage the biological integrity of surface waters. They provide an answer to the question, What concentration is too high for the many environmental contaminants of concern in water, sediment, and organism tissue? However, there are a number of problems and shortcomings with the approaches that are used today but that were developed 10 to 20 years ago. Though all water-quality protection programs have intended levels of protection, the exact level of protection provided by the criteria—that is, the probability that undesired events will occur—is unknown. Almost certainly, the level of protection of criteria varies from application to application. To some extent, this is due to the "one-size-fits-all" methodology used for their derivation.

There is no doubt that the adoption of a uniform methodology for criteria derivation was a great step forward, when compared to the unstructured procedures that it replaced. However, the need for uniformity imposed restrictions that are no longer necessary. Modern advances in toxicological and chemical understanding and modeling techniques can be used to provide greater flexibility and site-specific protection.

The issue of bioavailability is a case in point. In the original water-quality criteria, the total chemical concentration is used. Recent advances have indicated that only a fraction of the chemical species are the biologically active forms and that other constituents in the water affect the bioavailability of chemicals and hence the toxicity or bioaccumulation that results. For uniformly derived criteria, these effects are usually not considered. However, they have been shown to have important effects in both increasing and decreasing the toxicity of chemicals.

Another important shortcoming is that WQC are not currently available to assess mixtures of chemicals as they occur in the environment. Chemical properties that influence toxicity and fate in the environment (e.g., persistence, hydrophobicity, partitioning, and complexing behavior) have been identified, and in some cases, practical models are available to quantify the extent of their importance. These properties and models can be employed directly to deal with mixtures.

Cross-media consistency is also a problem. Water, sediment, and tissue criteria development methodologies need to be harmonized. That is, the level of protection afforded by these criteria should be the same, and undesired effects should not simply be shifted from one medium to another. Again, the necessary technical information to accomplish this is available for some classes of chemicals. Exposure can be assessed using models that relate concentrations among the various media allowing cross-media consistency. The ultimate goal is to provide aquatic life criteria that are protective of all aquatic ecosystem components.

The Workshop Focus

The primary question posed at the workshop was, if one were to design an aquatic-quality criteria methodology from scratch, so that as many as possible of the problems enumerated above are addressed and if possible solved, what would the criteria look like? The design should not be constrained by previous formulations, nor by historical precedents, nor by familiarity with present procedures. Rather it should use the knowledge and data that are currently available, or that can be reasonably assumed to be available in the near future.

Conclusions

A number of interesting conclusions arose from the workshop deliberation. The initial observation was that the conceptual model and general procedures used in the current derivation of WQC are reasonable. In a phrase, WQC are not too bad, although they are generally thought to be conservative in most cases.

However, in order to address the problems enumerated above and others presented in the subsequent chapters, it was decided that there should be 3 different types of criteria. These are designed to provide the means to achieve a specified level of protection with progressively less uncertainty. This is achieved by progressively considering more site-specific, chemical-specific, and organism-specific factors. These criteria are referred to as Type 1, 2, or 3 criteria. Although they do have an increasing level of sophistication and data requirements—they are progressively more chemical, site, and organism specific—they should not be thought of as tiered criteria.

Type 1 criteria

Type 1 criteria are what would currently be a country's national criteria. They are envisioned as broadly applicable to the protection of aquatic organisms. By design, they are protective, but not necessarily predictive, of environmental effects. They are also the least modified for site-specific applications and the most frequently used for water-quality assessment and permit writing. These criteria should protect aquatic life under most circumstances. Current Type 1 approaches have been very useful, have successfully protected aquatic life, and in many cases, have provided for the return of healthy aquatic communities to severely affected waters. Therefore, many characteristics of the existing criteria should be retained.

Type 2 criteria

Type 1 and 2 criteria are similar. The difference is that Type 2 criteria include methods to deal with the physical and biological characteristics of the site that are too complex to be incorporated in the broadly applicable Type 1 criteria. Considering these factors decreases the uncertainty in setting the criterion, without departing from the framework embodied in Type 1 criteria or altering the desired level of protection. Type 2 criteria include components that can be refined to greater site-specificity and relevance. These components can be classified into 2 broad categories: factors affecting chemical bioavailability at the site and factors affecting biological distributions at the site.

There is no firm division between Type 1 and 2 criteria. As methods are developed that can be broadly applied—for example, corrections for bioavailability that are based on chemical parameters that are straightforward to measure—they would be included in Type 1 criteria. That is, methods that would be state of the art today, and thus be part of Type 2 criteria, may become commonplace in 5 years and become part of Type 1 criteria.

Type 3 criteria

In contrast to Type 1 and 2 criteria, Type 3 criteria are quite different. Type 3 criteria are designed to more fully integrate risk-based, site-specific assessments. They involve a thorough examination of the resource to be protected, with emphases on its designated use (including asking if it is appropriate for the site), cross-media analyses and harmonization of criteria (e.g., air, water, sediment, tissue), and consideration of multiple lines of evidence (e.g., biological surveys, in-situ toxicity tests, habitat assessments). Type 3 criteria are required if

- unusual or difficult chemical fate and transport or bioaccumulation issues exist,

- unusual or difficult environmental toxicity issues (e.g., absence of suitable surrogate species for a unique water body such as the Great Salt Lake, absence of suitable wildlife toxicity data for a bioaccumulative chemical, etc.) are present,
- focus is needed on a specific population for which the current species protection approach is not appropriate,
- Type 1 or 2 criteria are suspected of being either over- or underprotective, or
- a site-specific management goal is desired.

The principal difference is that Type 3 criteria are entirely site specific. The analysis tools that are employed can be similar to those used in Type 1 and 2 criteria, but they are tailored to the site. For example, rather than a generic fate and transport model, a specific model would be built and utilized. Or, a specific biological endpoint would be used—for example, concentration in wading bird eggs—rather than the generally protective endpoints that are derived from probability distributions of laboratory toxicity tests.

Implementation—Can They Be Built?

Type 1 and 2 criteria are built from the same scientific principles. For many of the problems discussed below, practical solutions are either currently available or will be available in the near term. For example,

- bioavailable metal concentrations can be evaluated using chemical models that account for speciation and active site binding,
- equilibrium partitioning (EqP)-based sediment guidelines for metals are inherently mixture criteria, and
- tissue criteria for dioxin and dioxin-like chemicals are mixture criteria that are based on an equivalent potency normalization for the various dioxin, furan, and polychlorinated biphenyl (PCB) congeners.

These are discussed in more detail throughout this book.

Type 3 criteria begin by identifying the resource to be protected. The U.S. Environmental Protection Agency (USEPA) Risk Assessment Guidelines provide the organizing principles for their development. Examples include

- systematic organization of the problem, the exposure, the effects, and the risk characterization;
- probabilistic methods for evaluating seasonal effects, varying exposures, co-occurrence and organism behavior; and
- risk estimates across all potential exposure concentrations using dose responses rather than point estimates.

Major benefits of implementation

Changing the criteria methodology entails certain costs and dislocations, so there must be some compensating benefits. Among these are

- level of protection is understood, and can be adjusted;
- uncertainty is reduced, and is better understood;
- economic efficiency is promoted through appropriate levels of protection;
- benchmarks are provided against which monitoring data can be evaluated; and
- risk communication is made more effective.

A more detailed presentation of the ideas outlined above for the new criteria can be found in the chapters that follow. They are organized following the risk assessment framework, consistent with the proposal presented above.

CHAPTER 1

Problem Formulation

Thomas W. La Point, Scott E. Belanger, Trudie Crommentuijn,
John Goodrich-Mahoney, Robert A. Kent, Donald I. Mount, Douglas J. Spry,
Torgny Vigerstad, Dominic M. Di Toro, F. James Keating Jr, Mary C. Reiley

Statement of the Problem

Quantitative expressions of the characteristics or conditions that describe or define acceptable or desirable water quality are given names such as "water-quality criteria" (WQC) or "water-quality guidelines" (WQGs), among other specifically defined terms. For simplicity, this book uses the term "criteria" as a generic term for all of these values. Globally, environmental managers use criteria to protect and assess the ability of surface waters to support aquatic resources and designated or intended uses. Criteria often are used to regulate aqueous discharges of contaminants and serve as the basis for assessing attainment of beneficial uses. Criteria have been used over the past 3 decades to assess chemical safety and to protect aquatic environments from excessive releases of chemicals. In this regard, they have been a valuable tool for regulatory agencies and regulated facilities. Regulatory programs that use criteria are generally regarded as very successful in their efforts to protect aquatic systems. The success associated with the use of criteria has, to a large extent, been due to a conservative approach to their derivation.

Although criteria have proven useful and are thought to be protective, there are several limitations to existing approaches. Criteria are based primarily upon tests conducted with water column or epibenthic organisms, and the principal route of contaminant exposure is direct uptake from water. Most approaches currently do not incorporate exposure via the diet or from sediments. Different countries use different approaches and different endpoints that result in numerous sets of values for a variety of chemicals. These values can vary according to the particular organisms or ecosystems being protected, the level of conservatism and safety factors applied, and the statistical approaches used to assess the compiled exposure and effects data. In addition, they can involve loosely defined assessment endpoints, uncertain links between measurement and assessment endpoints, effects data with exposure conditions and pathways different from field conditions, and exposure evaluations that may be unrealistic or incompatible with laboratory effects evaluations. As such, the estimated risk to aquatic organisms when WQC

Reevaluation of the State of the Science for Water-Quality Criteria Development. Mary C. Reiley et al., editors.
©2003 Society of Environmental Toxicology and Chemistry (SETAC). ISBN 1-880611-30-9

are exceeded, and associated regulatory actions are often uncertain, gives little or no consideration to the issues of interactive effects (i.e., complex mixtures), bioaccumulative potential, multiexposure pathways (e.g., dietary, dermal), or mitigating factors. Moreover, current methods do not easily facilitate quantification of this uncertainty, nor do they provide for estimation of risks to higher levels of ecological organization (e.g., populations and communities). An integrated, holistic approach that considers multiple pathways for environmental exposure and that calculates the risk to aquatic life associated with a given concentration of a contaminant has been lacking.

In summary, the current WQC methodology is intended to be protective of aquatic organisms at the species, population, and community levels. It promotes the broadly accepted conclusion that they are protective under most circumstances, but have unknown levels of certainty. However, the approach is not designed to estimate the level of risk or protection afforded by the criterion for a given contaminant. The criteria only define a chemical concentration, in water, below which certain toxic effects should not be expected in representative organisms, or project a point estimate of the water concentration that would yield a protective level in edible fish tissue. As a result of applying current WQC, it is likely that some waters are adequately protected, some are overprotected, and some are underprotected, but the degree of protection provided is unknown.

There have been many advances in the science of environmental toxicology and chemistry, in the development of the current criteria, and in the consequences of implementing water-management-based criteria. The expectation of dischargers, regulators, and the public is that criteria are based on the best science currently available and that application of criteria does not result in wasted resources and arbitrary regulatory decisions. Criteria need to be improved, not to become more or less stringent, rather to be better integrated and more closely aligned to the attributes we wish to protect. A risk-based approach that considers exposure and effects parameters collectively holds the greatest promise.

Summary of Current Practice

> Criteria are the scientific data used to define a limit of variation
> or alteration of water quality judged to not have some specified,
> usually adverse, effect on the use of a water by man or organisms
> inhabiting the water.
>
>   — *Charles Warren, 1971*

A number of countries (New Zealand, Netherlands, Canada, U.S.) were represented in the workshop, and in each, the purpose of criteria (or guidelines) is to protect aquatic systems and their uses. Though the names, implementation, and enforcement approaches differ, all use the same general conceptual model to derive criteria. This model requires a minimum set of laboratory-generated toxicity data for aquatic species, and these data are extrapolated or interpolated to derive a criterion, that is, a chemical concentration limit thought to be protective of those species. This model assumes that criteria estimated from a subset of species will be reasonably protective of the majority of species, both tested and untested, and that species protection confers ecosystem protection. For example, in the U.S., aquatic life criteria are derived from a minimum database that includes acute toxicity tests for at least 8 representative genera and chronic toxicity tests for at least 3 genera. The criteria are the concentrations of a chemical in water estimated to protect 95% of genera that are similar to those tested. Allowances for duration of exposure and frequency of exceedance also are specified (Stephan et al. 1985).

A more detailed review of approaches used in the participating countries is presented in Appendix 1.

Specific problems with the current practice

All of the participant countries share, in varying degrees, 3 technical problems with the way criteria are derived and implemented. The problems are described briefly below and are the subject of the balance of this book.

1: Criteria derivation

The first problem can be succinctly stated as "one-size does not fit all." Typically, criteria are developed from laboratory data in a uniform way for all chemicals. Minimal consideration is given to such concerns as mode of action (MOA), critical body residues (CBRs), chemical mixtures, dietary exposure, environmental chemistry, or population effects. With some exceptions (e.g., certain metals and ammonia), criteria are applied to all types of water bodies without regard to site-specific features, chemical or biological, that influence and modify toxicity of the chemical.

2: Criteria implementation

The second problem relates to the manner in which criteria are used, specifically the averaging period (duration of exposure of organisms to the chemical) and the allowable frequency of exceedance (how often the criterion can be exceeded without appreciable harm to population recovery). To the extent that these have been included in some criteria, they also suffer from the "one-size does not fit all" problem. Both are presently seen as constants for all chemicals and locations. Another implementation issue is the area to which criteria should apply; how large or small an area of a system should be covered by any given criterion? Given the heterogeneity of ecosystems, the more site-specific a criterion is, the more likely it is appropriate for the given area.

3: Cross-media harmonization and multipathway exposure

The third problem is a need for consistency in developing and implementing criteria among media. Species in the water column or sediment and wildlife and human consumers should all be protected to the same extent. Cross-media harmonization is attempted in all countries (e.g., the Dutch Ministry of Housing, Physical Planning and Environment [VROM] 1999), but with limited success, because the scientific basis of these procedures needs to be improved. Little consideration is given to multipathway exposure, leaving criteria to reflect uptake from water but not from diet or dermal contact.

Problem Resolution

There has been substantial progress in toxicology, ecology, chemistry, and statistics that directly affects the degree to which we can resolve the scientific aspects of these problems. Chemicals can now be grouped into useful classes according to mode of toxic action and environmental behavior. Chemical properties that influence toxicity and fate in the environment (e.g., persistence, hydrophobicity, partitioning, and complexing behavior) have been identified, and for some chemical classes, practical models are available to quantify the extent of their importance.

Flexible computer models coupled with powerful personal computers can now significantly enhance implementation procedures, such as exposure assessments. Probabilistic models, used to compute variability in exposure, can compute exceedance frequencies of criteria for various averaging periods, simultaneously including the variability of other factors such as streamflow and pH. Cross-media consistency can be improved using models or data that relate concentrations among the various media.

Specific scientific issues that must be considered or addressed

A number of issues were identified during the workshop because of known scientific weaknesses about the development and implementation of criteria or because the concepts have been inappropriately incorporated into the current criteria process. Those discussed during the workshop are listed below:

- Different classes of compounds require different data and treatment according to their MOA and chemical properties.
- Regression analysis (ECx [effect concentration for x% of the population] rather than hypothesis testing (no-observed-effect concentration [NOEC] or lowest-observed-effect concentration [LOEC]) is needed to properly express chronic effects.
- Background concentrations, essential elements, and diet should be considered in both toxicity testing and criteria setting.

- Effects are highly dependent on bioavailability for many classes of compounds; bioavailability varies with site-specific characteristics that are not well understood.
- The definition of "exceedance frequency" is a large source of conservatism when implementing the criteria; research is needed on factors that influence appropriate exceedance frequency.
- Chronic data should be used in place of acute-to-chronic ratios (ACRs) to develop chronic criteria.
- The assumption that criteria that protect 95% of the tested genera protect entire aquatic ecosystems needs to be tested.
- The toxicity response distribution needs to be extrapolated properly.
- Uncertainty in criteria needs to be quantified.
- Methods to harmonize criteria across media are needed.
- Averaging periods should better match the biological response period associated with the toxicological endpoint.
- Implementation procedures (e.g., for calculating permit limits) should not add unknown margins of safety.
- The range of biota to be explicitly protected should include amphibians and wildlife.

Conceptual Model Development

In ecological risk assessments (ERAs), the process of problem formulation helps to define the nature and extent of the problem, the resources at risk, the ecosystem components to be protected, and the need for exposure and effects data. The product of the problem formulation process is the conceptual model that describes critical exposure pathways and stressor-response links that impact the resources at risk. The first step in problem formulation, and the creation of a conceptual model, is to identify the features of the ecosystem and resources to be protected, the key stressors, their characteristics, and the adverse ecological effects that could result if they are not addressed. The second step is to select assessment and measurement endpoints. Assessment endpoints are expressions of attributes that we wish to maintain or promote and that relate the stressor-induced effects to desired use of the water body. Measurement endpoints refer to the specific data collected and are evaluated to determine whether the assessment endpoint is achieved. The final step is to assemble all these elements in the conceptual model.

Problem formulation, as applied to criteria development, identified 3 facts:
1) characteristics of chemical stressors (e.g., speciation, partitioning) vary across ecosystems, 2) the structure and function of communities and populations are different across ecosystems, and 3) critical effects or responses vary across ecosys-

tems. For these reasons, many measurement endpoints linked to the assessment endpoints need to be site-specific in nature, although the assessment endpoints may remain constant. Therefore, the conceptual model for deriving criteria must acknowledge the need for different types of criteria to account for variation within and among different aquatic ecosystems.

Conceptual model: Three types of criteria

There are 3 types of criteria proposed as alternatives to existing WQC approaches, each of which may be based on a somewhat different conceptual model. Type 1 "national" or "broadly applicable" criteria are developed based upon conservative assumptions and application across a variety of water types, water quality, and biological conditions, just as the current criteria are derived. The conceptual model for WQC traditionally has been based on an assessment of the effects of chemicals in water on fishes, invertebrates, and algae. This workshop recommends that this approach be expanded to include pathways such as exposure to contaminated sediments and/or exposure through diets in the derivation of a Type 1 criterion. In addition, Type 1 criteria should account for as many site-specific factors as possible. To the extent that generic environmental factors may consistently affect criteria, such relationships may be used to incrementally reduce the conservative nature of Type 1 criteria. These include site-specific factors that are ubiquitous, simple, and inexpensive to measure, such as pH, dissolved organic carbon (DOC), hardness, and salinity. Type 1 criteria may be expressed as narrative statements, numbers, formulas, or models (e.g., exposure, bioavailability, bioaccumulation, toxicokinetics).

For circumstances in which Type 1 criteria are known to be inaccurate because of site-specific conditions or have a high degree of associated uncertainty, Type 2 criteria should be derived. Type 1 and Type 2 criteria are based on essentially the same conceptual model. The difference between the two is the level of the site-specific information used to derive the criteria. Site-specific factors may include more detailed field and laboratory measures of chemical bioavailability (e.g., a water–effect ratio (WER) or field-derived bioaccumulation factor [BAF]), recalculation based on presence or absence of sensitive species, and identification of habitat issues (nonchemical stressors). Type 2 criteria derivation may result in a criterion that is greater than or less than a Type 1 criterion, but will be more certain and more accurate.

Type 3 criteria will more fully integrate risk-based, site-specific assessments. They involve a thorough examination of resources to be protected, with emphases on designated uses, cross-media analyses, harmonization of different criteria (e.g., air, water, sediment, tissue), and consideration of multiple lines of evidence (e.g., biological surveys, in situ toxicity tests, habitat assessments). Type 3 criteria are reserved for cases in which the general conceptual model, on a national or site-specific scale, simply does not apply. Type 3 criteria also are appropriate where there is convincing evidence that the Type 1 or 2 criteria are underprotective or overprotective or there is a site-specific issue not considered by Type 1 or 2 approaches (e.g., endangered or threatened species). Development of Type 3 criteria typically requires considerable site-specific data, community involvement, and peer review in their derivation.

In the progression from Type 1 to Type 2 to Type 3 criteria, the degree of accuracy in achieving the desired level of protection increases while the amount of uncertainty decreases. As depicted in Figure 1-1, Type 1 criteria are the most generically applied and the most immediately available, whereas Type 3 criteria require the most technical sophistication and are the most site-specific. Figure 1-2 depicts the application of the 3 criteria types to hypothetical sites in relation to theoretically known "correct" effect concentrations. In Figure 1-2A, a generically applied Type 1 criterion is often overprotective to varying degrees, but occasionally underprotective. In Figures 1-2B and 1-2C, Type 2 and Type 3 criteria move closer to the "correct" concentration because these types are more site and problem specific.

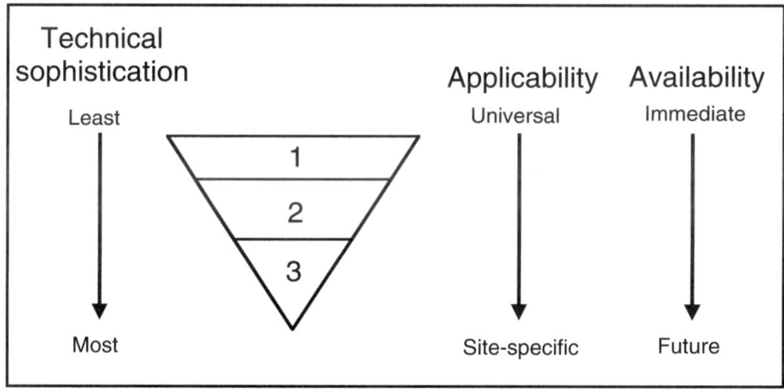

Figure 1-1 Relationship of technical sophistication, criterion applicability, and availability to assessment strategy

Type 1 criteria

Type 1 criteria are envisioned to be broadly applicable "national" criteria developed based on very simplified elements of ERA. Type 1 criteria are analogous to the current ambient water-quality criteria (AWQC) in the U.S., environmental-quality guidelines in Canada, maximum permissible concentrations (MPCs) in the Netherlands, and WQGs in New Zealand and Australia. Type 1 criteria may be derived for acute and chronic exposures. For example, in the U.S. Environmental Protection Agency (USEPA) procedure for deriving AWQC, criteria maximum concentrations (CMC), and criteria continuous concentrations (CCC) are derived

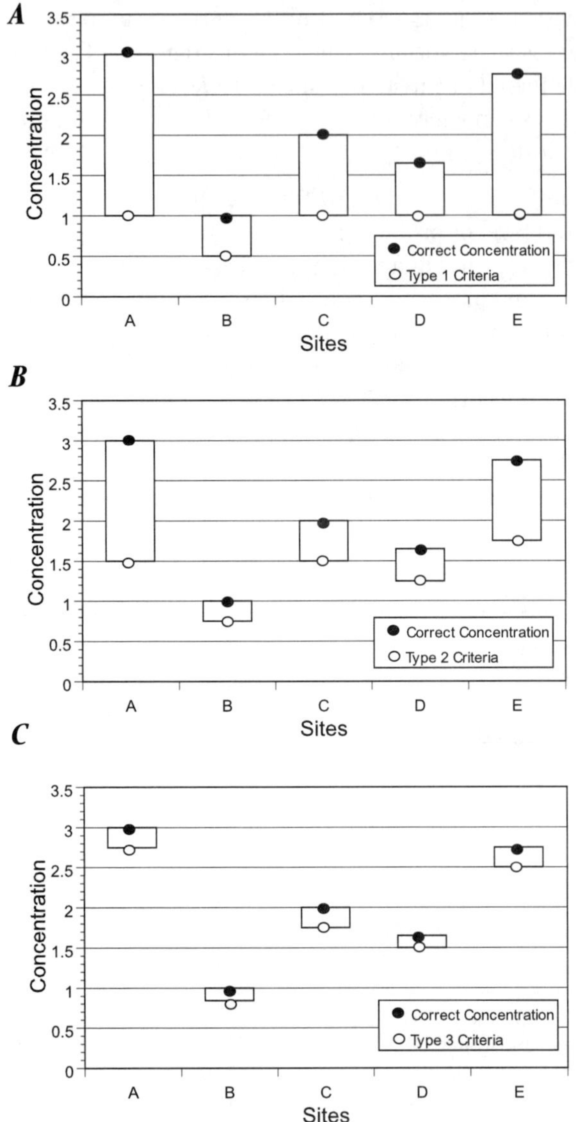

Figure 1-2 The 3 types of criteria provide different degrees of allowable uncertainty and expected accuracy

for acute and chronic exposure conditions, respectively. As national criteria, they are broadly applicable for the protection of aquatic organisms. By design, they are protective, but not necessarily predictive, of environmental effects. They also are the least modified for site-specific applications and the most frequently used for water-quality assessment and permit writing. These criteria should protect aquatic life under most circumstances. Based on collective experiences, workshop participants generally believed that current approaches have been very useful, have successfully protected aquatic life, and, in many cases, have provided for the return of healthy aquatic communities to severely affected waters. As such, participants concluded that many characteristics of the current criteria derivation approaches should be retained.

Strengths of the current approach

The workshop participants agreed that there are several strengths of the current approaches. For example, most countries use similar toxicity tests and generate similar types of data. There is a minimum data set that provides confidence that most aquatic biota will be protected. Current approaches consider some aspects of toxic chemical bioavailability, for example, water hardness effects on heavy metal availability and uptake. There also is some provision for assessing bioaccumulative chemicals through criteria for tissue residue values estimated from bioconcentration factors (BCFs). Current approaches used to derive criteria that are broadly applicable cost substantially less to establish than criteria developed on a site-by-site basis where there are significant requirements for site-specific information collection. This is especially efficient in cases where minimal data are developed and there are few issues associated with criteria compliance. Once established, however, those same criteria may lead to great inefficiency if unnecessary compliance costs or limited resource management efforts are wasted because there is insufficient recognition of the need (or insufficient opportunity) to adjust criteria to more accurate and more certain values.

Areas for improvement

Type 1 criteria should provide as much site-specific information as possible, while ensuring broad and general applicability, and without sacrificing confidence in the protectiveness of the criteria. For example, current criteria could better account for factors influencing bioavailability. Chemical modeling may take the place of more expensive and time-consuming, site-specific bioassays. Effects of pH, DOC, or suspended solids on bioavailability could be modeled, as hardness is now.

Different chemical classes (e.g., metals, organometals, poorly soluble organics, surfactants) also may have unique environmental characteristics or different MOAs. These different chemical classes may require unique methods for deriving criteria. Newer endpoints, such as toxicity equivalents for dioxins or bioindicators of significant effects, might be added to traditional measures of survival, reproduction, and growth. Acute and chronic criteria may not be needed for all chemicals, for example, where the ACR is close to 1.

Chemical concentrations in the natural environment vary considerably over time and space, whereas criteria are based on laboratory bioassays using constant concentrations. This means that, in most cases, the criteria are not matched to actual exposure in the field. Certain provisions already in the U.S. guidance are not often used, perhaps because better guidance is required. For example, there are provisions for considering plant data and tissue residues for human and wildlife consumers of fish that should be used more often than they are now. Existing WQC often are used as stand-alone standards rather than considering potential cross-media transfers. With a better understanding of relationships among different media—air, water, sediment, and tissue residues—there is an opportunity to derive appropriately protective criteria for each of these media. Criteria specific to the medium of exposure and type of protection (e.g., sediment-quality guidelines (SQGs) and tissue residue guidelines) are more scientifically valid than are values expressed as aqueous concentrations.

If there is evidence that a Type 1 criterion is incorrect for a particular site or region (e.g., no evidence of effects in areas where exceedances are common), or there is considerable uncertainty about whether the criterion is correct (e.g., small effects database, likely sensitive species underrepresented or overrepresented in effects database, physical–chemical properties at a site that could significantly alter toxicity), then consideration should be given to developing a Type 2 criterion.

Type 2 criteria

The difference between the conceptual models used for Type 1 and 2 criteria is the degree to which properties of the site are taken into consideration in the effects and exposure characterization. The intent is for Type 2 methods to deal with the physical and biological characteristics of the site that are too complex to be incorporated in Type 1 environmental-quality criteria. Considering these factors decreases the uncertainty in setting the criterion, without departing from the framework embodied in Type 1 or altering the desired level of protection. Type 2 criteria include components that can be refined to greater site-specificity and relevance. These components can be classified into 2 broad categories:

1) factors that affect bioavailability and toxicity of chemicals at the site, and

2) factors that affect the numbers and types of organisms at the site.

Type 2 criteria may consider the sensitivity of the species expected at the specific site, the effect of site environmental conditions on toxicity of the substance, or the effect of both species sensitivity and water quality.

Chemistry, fate, and bioavailability

Bioavailability is a function of the exposure scenario in aquatic ecosystems and, thus, of how the chemical interacts with the organism. Bioavailability is usually taken to be the proportion of chemical that is taken up by the organism and transported to a target receptor. The amount of chemical actually available for uptake by organisms in the natural environment is usually less than that available under traditional laboratory conditions designed to measure toxic effects. Numerous modifying factors in the environment lend themselves to reducing the bioavailability of a given chemical species—via adsorption, complexation, or transformation. When these factors are understood and incorporated into Type 2 criteria, the manager is able to obtain a much better estimate of potential effect concentrations with a greater degree of confidence.

Understanding chemical behavior (complexation, speciation, physical or chemical and biological degradation) is critical to understanding how bioavailability is affected by site-specific water quality. For example, metal toxicity is highly influenced by complexation with DOC, aqueous ionic constituents (Ca^{2+}, Mg^{2+}, Cl^{2-}), and suspended solids. Silver is an excellent example of a metal that strongly binds with anionic compounds and in natural waters tends to be less toxic than in the lab. These factors may be accounted for by conducting laboratory tests with site water or by utilizing water-quality models that account for biological and chemical interactions.

Biological considerations

Several approaches already exist to modify Type 1 criteria based on biological considerations. A criterion value may be made more appropriate to a site by deleting species or life stages not expected to occur at the site, or by adding species or life stages that better represent those found at the site. After the toxicity database has been amended, the criterion is recalculated. This procedure may require additional testing to meet minimum data requirements. The database on the sensitivities of species expected to occur at the site can be compared to the measured or predicted concentrations in the medium of interest to conceptualize species and functions potentially at risk. Another way to calculate a Type 2 criterion is to test resident species (those species that actually reside, or should reside, at the site under investigation) in site water and use the results to derive a new criterion. This requires generating a separate set of toxicity data for the site water to use in place of the laboratory-generated data set.

For highly bioaccumulative substances, Type 2 tissue residue criteria also may be derived for conditions at the site that best reflect local conditions of trophic composition and food web patterns. This may require a site-specific, field-derived BAF or biota–sediment accumulation factor (BSAF). These data should be obtained for receptor species relevant to the site, as opposed to using default values and taxa considered in the national criteria. Experience has shown that with highly bioaccumulative compounds, the organisms of primary concern are at the top end of the food web. Insectivorous and piscivorous birds are known to be at risk for compounds such as selenium, methylmercury, polychlorinated biphenyls (PCBs), and dioxins.

Type 3 criteria

If for a particular water-quality management issue there is concern about the conceptual model validity or level of uncertainty about a Type 2 criterion, then there may be a need to develop a Type 3 criterion. In Type 3 criteria, site-specific evaluations of biological and physical factors, microcosm and mesocosm test results, and field data may be used to develop a criterion that has a low level of uncertainty while achieving the desired level of protection.

Type 3 criteria are desirable when unusual and significant scientific, socioeconomic, or technological considerations prevail at the site, and when stakeholders agree that the limited ERA conceptual model used for Type 1 or Type 2 criteria does not provide an appropriate outcome. Type 3 criteria are "issue driven" and consider the problem from the top down. In contrast, Type 1 and 2 criteria are focused on single chemicals and take a bottom-up approach. Type 3 criteria should become increasingly predictive of water-quality effects thresholds while maintaining a suitable level of protection. A Type 3 criterion may be appropriate when data on any of the following are available and usable:

- locations of important and sensitive habitats in the water body;
- seasonal variations in water quality, as they relate to use of the water body by species or life stages;
- site-specific issues affecting the ability of biota to recover following elevated exposure to a stressor;
- risks posed by the environmental-quality parameter for which the criterion is being developed are small when compared to risks posed by other noncontaminant ecological stressors;
- unusual or difficult chemical fate and transport or bioaccumulation issues;
- unusual or difficult environmental toxicity issues (e.g., absence of suitable surrogate species for a unique water body such as the Great Salt Lake, or absence of suitable wildlife toxicity data for a bioaccumulative chemical);
- population-level or community-level risks;
- risks created by actions taken to meet a criterion (e.g., habitat loss following dredging or capping of contaminated sediments, hydrodynamic changes due to modifying a point source discharge, or controlling runoff or groundwater discharge);
- site-specific management goal; and
- risks to threatened or endangered species.

Use of comprehensive ERA
The recommended tool for Type 3 criteria is a full, quantitative, site-specific risk assessment. In contrast, Type 1 and Type 2 criteria rely on a variety of default values; well-defined models; limited, site-specific data; and considerations mostly driven by toxicity data and assumptions about exposure. ERA is a normative framework for evaluating the impacts of environmental conditions on organisms, populations, or communities. Many agencies and organizations have endorsed the concept of ERA. For example, the USEPA produced a widely cited set of guidelines for ERA (USEPA 1995). The USEPA guidelines are a framework, not a detailed guidance about how to conduct an ERA. The details of risk assessment are gener-

ally worked out on a case-by-case basis. As the field matures and experience grows, there will likely be more standardization of methods and data.

Ecological risk assessment, using the terms defined in the USEPA guidelines, is composed of 3 primary phases: 1) problem formulation, 2) analysis, and 3) risk characterization. Problem formulation is the most important phase in risk assessment, in the sense that if the problem is ill structured, the analysis and risk characterization phases will be wasted efforts. Some key steps of problem formulation are

- establishing the management goal for which risk assessment is being conducted;
- determining what will be assessed to evaluate whether the management goal is being met and, if not, how to meet it;
- defining specific measures of exposure and effect to be used as surrogates for the assessment endpoints (which are not always measurable);
- developing a conceptual model of how stressors in the environment reach the measurement endpoints; and
- developing an analysis plan to collect the exposure and effects data needed for the analysis phase of the risk assessment.

The analysis phase includes exposure assessment and effects characterization. Exposure assessment estimates the levels of stressors to which receptors are exposed. To develop Type 3 WQC for a substance, there are 2 major components of exposure estimation: 1) measuring or predicting environmental concentrations of the compound in each environmental medium through which a receptor may be exposed, and 2) computing average spatial and temporal exposure concentrations. Effects characterization estimates the threshold level of exposure to a substance for unacceptable effects. This may be a time-variable estimate, but typically the duration of exposure is assumed when estimating effects thresholds.

The final stage of the risk assessment process is risk characterization, an interpretive phase consisting of risk estimation and interpretation. Risk characterization combines exposure analysis and effects analysis. In a very simple approach, this can be expressed as a ratio of exposure concentration to effects threshold. In a more sophisticated approach, the probability of exceeding an effects threshold in a population of organisms or species can be computed. Risk interpretation entails evaluating risk estimates in the context of all available exposure and effects data, as well as additional lines of evidence, such as ambient bioassays, biological surveys, and histological studies.

Summary of criteria type selection factors

For a given chemical and site, resource managers must decide what criterion type is most appropriate. This decision must be based on information available for the chemical and site in question. Factors may include the level of regulatory concern about potential adverse effects or the degree to which a permittee finds a national criterion to be more restrictive than necessary to support a designated use of the water body.

Type 1 criteria establish the minimum data and models that must be available, or easily obtained, to calculate a national value. It should be possible to derive Type 1 criteria for any chemical of concern using standardized toxicity test results and default equations and model inputs (e.g., hardness, pH, DOC, BAF, BSAFs). Type 1 criteria are broadly applicable and ensure that the most sensitive species are protected most of the time. Because they are broadly applicable, they may be appropriately protective, underprotective, or overprotective of a particular site because of local water-quality conditions, discharge characteristics, or sensitivity of resident biota.

The impetus to develop a Type 2 criterion for any given chemical may come from a number of needs, such as data or model availability, discharge permit development process, or because the chemical is of significant concern. A Type 2 criterion should be more accurate than a Type 1 criterion for a site, given the particular characteristics of a discharge or receiving water.

In cases where the limitations of a Type 1 and Type 2 criterion do not appropriately address site complexities, then a Type 3 criterion should be developed. A Type 3 criterion should use the full potential of ERA. Potential complexities include unique endpoints with undefined relationships to the standard test organisms used to develop Type 1 and 2 criteria (e.g., ensuring that the criterion protects sensitive and endangered species). Type 3 criteria should provide highly accurate and site-specific criteria with a specified level of protection designed to protect all components of the local aquatic ecosystem on which it is based.

References

Stephan CE, Mount DI, Hansen DJ, Gentile JH, Chapman GA, Brungs WA. 1985. Guidelines for deriving numerical national water quality criteria for the protection of aquatic organisms and their use. Washington DC, USA: USEPA. NTIS PB 85-227049.

[USEPA] U.S. Environmental Protection Agency. 1995. Guidelines for ecological risk assessment. Washington DC, USA: USEPA. EPA-630-R-95-002F.

[VROM] Dutch Ministry of Housing, Physical Planning and Environment. 1999. Environmental quality objectives in the Netherlands. NL: VROM.

Warren CE. 1971. Biology and water pollution control. Philadelphia PA, USA: W.B. Saunders. 17 p.

CHAPTER 2

Exposure Analysis

William H. Benson, Herbert E. Allen, John P. Connolly, Charles G. Delos,
Lenwood W. Hall Jr, Samuel N. Luoma, David Maschwitz, Joseph S. Meyer,
John W. Nichols, William A. Stubblefield

Introduction

This chapter presents the state of the science regarding the evaluation of exposure as it relates to WQC, SQGs, and wildlife criteria (WC). Throughout this discussion, attempts are made to identify the methods that are currently available, the limitations to the methods, and the possible solutions to the limitations. Strategies that permit a more accurate reflection of the true nature of exposure are identified. Finally, research needs are identified and prioritized with regard to necessity and likelihood for success.

At present, WQC are derived from an empirical approach founded on single-species laboratory toxicity tests with little consideration given to exposure parameters that may affect the amount of chemical that actually reaches the site of toxic action. As such, the current approach for development of WQC is open to criticism. A mechanistic approach would result in increased accuracy of the exposure description, increased flexibility to consider complex exposure scenarios, and increased predictive capability through optimization of exposure models. Therefore, the exposure group focused on increasing the sophistication of current approaches to exposure analysis. This mechanistic approach should also improve exposure estimates used in deriving SQG and WC.

Workgroup participants recognized that a hierarchical relationship best describes the process for WQC development and implementation (Figure 2-1). This approach is equally valid with respect to the technical sophistication of criteria development, as well as its application and uses. Likewise, the availability of methods and procedures used to derive criteria range from currently available, through near-term (3 to 5 years), to long-term developments (5+ years).

The Exposure Analysis Workgroup addressed procedures to describe and predict the exposure concentrations needed to assess toxicological effects. The topics discussed included spatial and temporal variations in toxicant concentrations, chemical partitioning, and speciation as they influence bioavailability. In addition, other chemical and physical characteristics that might modify toxicity were considered. Multimedia interactions were discussed with respect to harmonization

---

Reevaluation of the State of the Science for Water-Quality Criteria Development. Mary C. Reiley et al., editors.
©2003 Society of Environmental Toxicology and Chemistry (SETAC). ISBN 1-880611-30-9

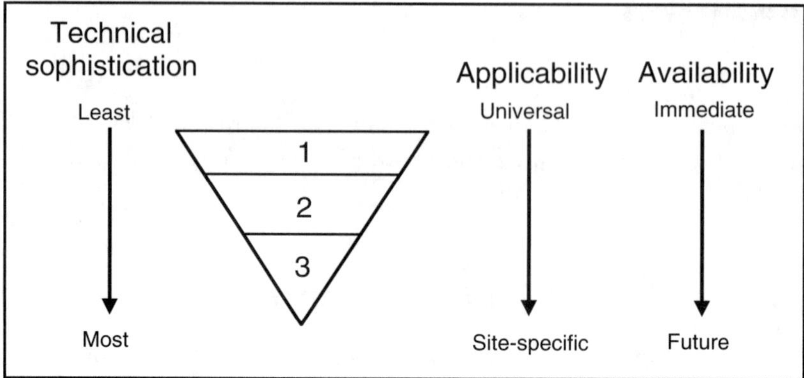

Figure 2-1 Relationship of technical sophistication, applicability, and availability to tiered assessment strategy

of water, sediment, and wildlife criteria or guidelines, as well as the integration of chemical exposures resulting from water, air, and soil. Finally, multipathway exposures were discussed with respect to their importance in the criteria or guidelines-setting process. Specifically, the group addressed the significance of dietary routes of exposure and evaluated the utility of tissue residue-based approaches as tools for use in the criteria development process.

"Exposure," as defined by the workgroup, is the co-occurrence of a chemical and a biotic receptor when there exists a realistic pathway of interaction. WQC are currently developed on a single-chemical basis, while "real-world" exposures typically occur in the form of chemical mixtures and modifying environmental factors. The influence of multichemical exposures is not considered in the present criteria derivation process. However, methodologies and procedures can be adopted from risk assessment practitioners for assessing the toxicity of chemical mixtures, and these approaches should be considered in the criteria implementation process.

With the exception of the metabolites of dichloro diphenyl trichloroethane (DDT), current WQC in the U.S. are based only on the occurrence and toxicity of the parent chemical substance. However, in the not too distant future, biotic and abiotic transformation processes might be sufficiently understood to permit estimation of transformation products and to deserve consideration in the criteria derivation and implementation process.

In summary, the exposure workgroup concluded the following:

- There should be a greater use of mechanistic models to predict bioavailability and chemical effects among different sites as a function of water-quality parameters, the chemical form in water, other water constituents, and the interaction of the contaminant with the biotic receptor site.

- Exposure variability should be recognized by deriving chemical-specific averaging periods from kinetic-based toxicity and accumulation models. These models should include a consideration of rates of recovery from catastrophic stresses, mechanisms of action, and effects on particular life stages of taxa with different life history strategies and degree of toxicant sensitivity. Appropriate mixing zones for chemicals should be derived from the kinetics of toxicity, behavior of receiving water organisms, and concentration isopleths associated with specific acute and chronic effects.

- Uncertainties in developing and applying criteria associated with multipathway exposure should be recognized through limits on tissue residues. A comprehensive understanding of dietary exposure should be developed for use in quantitative bioaccumulation models, using output from such models to replace the BAF concept. A dose–effect database also should be developed that incorporates toxicokinetic data and that expresses toxicity in terms of both exposure concentration and the chemical concentration at the site of toxic action or in an appropriate surrogate tissue. Estimates of dose to target tissues should be included in new bioassay protocols.

- Media-specific criteria should be harmonized by coupling existing media-specific models and conducting demonstration projects at well-characterized field sites. Spatial and temporal scales of exposure for chemicals lacking environmental exposure data should be characterized by monitoring appropriate media at scales of exposure typical of contaminated areas.

- A comprehensive, hierarchical understanding of biological responses to pollutants is needed to link chemical exposure to responses of the receptor site, organisms, populations, communities, and ecosystems.

Bioavailability

Current approach for WQC development

Present WQC do not, in general, consider bioavailability of chemicals. The majority of chemical criteria are expressed on a total chemical basis with no modifying factors considered. Metal criteria were originally developed and applied on a total metal basis; however, current USEPA guidelines recommend comparing dissolved metal concentrations in receiving waters to the criteria (Prothro 1993). This change reflects an understanding that the dissolved metal concentration more closely approximates the bioavailable fraction than does the total metal concentration. The few criteria that incorporate modifying factors use empirically derived relationships rather than mechanistic considerations. In fact, the only modifying factors that are incorporated are pH for ammonia and pentachlorophenol and hardness for several metals in fresh water. Oxidation state is considered for arsenic and chromium, and the chemical species mercury(II) is distinguished from methyl mercury. For some hydrophobic organic compounds, the criteria developed under the Great Lakes Initiative considered the influence that dissolved organic matter (DOM) has on bioaccumulation.

Limitations of the current approach to WQC

A major limitation of most current WQC is their lack of consideration of the many factors that can modify bioavailability and reduce toxicity. In the U.S., and in some other countries, modifying factors are incorporated in WQC for several contaminants, for example, ammonia and some metals. However, in some cases, factors that were considered are less important than those that were not considered. For example, the WQC for copper includes an equation to adjust for the effect of hardness on observed toxicity, but the effect of pH on copper toxicity is much greater (Erickson et al. 1996). For hydrophobic organic compounds, no modifying factors are incorporated into the WQC (except as noted previously for the Great Lakes), even though several studies have shown that as DOM concentration increases, bioaccumulation and toxicity of a hydrophobic compound decreases (Landrum et al. 1987).

The present acute and chronic WQC for selenium are expressed as total elemental selenium. Several oxidation states of selenium exist in the environment. The predominant states are selenate (Se VI) and selenite (Se IV) in water and Se(0) and Se(−II) in sediments, each of which varies in terms of relative concentrations in specific water bodies and each having different toxic potencies (USEPA 1996; Canton 1999). The major effect of selenium results from the dietary uptake of organoselenium by waterfowl and fish, resulting in maternal transfer and teratogenicity (Adams et al. 1998). The rate of transformation of selenate and selenite to reduced and organo-forms depends on the interaction of a number of environmental factors including redox potential, algal and microbial productivity, and hydraulic retention time. Hence, the proportion of the total selenium that is in the more toxic form will depend on site characteristics (Adams et al. 1998). Consequently, a selenium criterion must incorporate selenium speciation to have general applicability.

The results of any toxicity test reflect the bioavailability of the test material in the test system. WQC are based predominantly on toxicity tests conducted by the addition of the toxicant either to a high-quality natural water or to a reconstituted water, and the chemical constituents of these waters interact with the added toxicant to modify its bioavailability. Therefore, to apply a criterion to a receiving water, the criterion must consider the bioavailability of the compound, both in the water used for criterion development and in the receiving water. Consideration of the bioavailability in the receiving water alone is not sufficient.

General mechanisms affecting bioavailability are modification of chemical forms or species or modification of the organism's interaction with the toxicant. One major mechanism that influences the form and bioavailability of most substances is partitioning to solids suspended in the water column or precipitated to bottom sediments. For acute toxicity in waterborne exposures, particulate-bound chemicals are not sufficiently bioavailable to contribute to observed effects.

Other factors that directly influence the chemical form of metals in freshwater include hydroxide ion, alkalinity, DOM, and to a lesser extent, other inorganic ions such as chloride and sulfate. In addition to the direct effects of these ligands on speciation, indirect effects must be considered. Calcium, magnesium, and hydrogen ions react with ligand sites on DOM to compete with metal binding to DOM, thus increasing the bioavailability of the metal. Similarly, hydrogen, calcium, magnesium, sodium, and potassium ions interact with ligands on gill surfaces to modify the organism's interaction with metals, bioaccumulation rates, and toxicity.

Steemann-Nielsen and Wium-Andersen (1970) first showed that the chemical form of copper controlled the observed toxicity. Most of the copper in seawater was nontoxic, and in recently upwelled seawater, phytoplankton growth was enhanced by the addition of a chelator. Sunda and Guillard (1976) showed that the toxicity of copper to phytoplankton could be correlated to the activity of cupric ion (Cu^{2+}) and not to the total copper concentration. Other investigators have reported that chemical species other than free, ionic metals are toxic (critical review by Campbell 1995). Guy and Kean (1980) and Borgmann and Ralph (1983) found that complexes of copper with synthetic ligands were toxic, but at reduced levels compared to free copper ion. Both Meador (1991) and Tubbing et al. (1994) reported that copper complexed with natural organic matter (OM) contributed to toxicity.

The principal factor that influences the bioavailability of hydrophobic organic contaminants is DOM. As the concentration of DOM increases, toxicity and bioaccumulation of a hydrophobic compound decrease (Landrum et al. 1987). The extent of interaction with DOM is a function of the compound's octanol–water partition coefficient (K_{ow}). Low K_{ow} compounds will be freely dissolved, but high K_{ow} compounds may be associated principally with DOM. A major factor in the speciation of ionizable organic compounds such as phenols is pH. At pHs where the compound is in the unionized form, interaction with DOM will be governed by K_{ow}; in contrast, the ionized form will be more water soluble.

Approaches to overcome limitations of current WQC

Near-term solutions

One approach to dealing with the issue of bioavailability is to incorporate a chemical speciation algorithm into the implementation of the criteria for a chemical. This approach assumes that a chemical species being considered can be correlated with the biological response. For example, acute and chronic cyanide criteria are based on measurements of acid-dissociable cyanide, which includes free cyanide and weak metal-cyanide complexes, and excludes strong metal-cyanide complexes such as ferricyanide. In this case, the analytical methodology includes the preponderance of cyanide that is capable of exerting the biological effect, although specific cyanide forms are not identified. However, in many other cases, notably metals, an analytical chemistry result cannot be correlated with the biological response across a wide range of water-quality conditions.

A single numerical value, even when modified by water hardness, does not provide the predictability of aquatic life effects for metals that is desirable for national criteria. Other factors, particularly pH and OM, modify the availability of copper and other metals to aquatic organisms (Erickson et al. 1996). Site-specific modification of the criteria is necessary to enable prediction of effects at a specific location. Currently, the WER is the recommended empirical method in the U.S. to provide site-specific modified criteria (Stephan et al. 1985; USEPA 1992, 1994b; Prothro 1993). To establish a WER, acute toxicity tests are conducted in a site water and in a reference (laboratory) water. Reference water tests are used as surrogates for the laboratory tests that were used to derive national criteria. The ratio of the effect concentrations is used as a multiplier to adjust the national WQC to account for differences in bioavailability, as measured by toxicity tests that would be applicable to that site. For example,

$$\text{Site-specific WQC} = \text{national WQC} \times \text{WER} = \text{national WQC} \times \frac{\text{LC50 site}}{\text{LC50 laboratory}} \qquad (2\text{-}1),$$

where LC50 is the lethal concentration for 50% of the population.

Brungs (1991) reported WER values for copper as large as 15. The WER value for copper was more than 5-fold greater than that for zinc at the same location, reflecting the stronger affinity of copper, relative to zinc, for organic substances.

Allen and Hansen (1996) analyzed the WER procedure in terms of the change in speciation that occurs as metal is added to a sample. Often, the molar concentration of OM in a natural water is not in vast excess of molar metal concentrations, particularly the concentrations of metals used in toxicity tests. Therefore, the fraction of the metal that is bioavailable will increase with added metal. This implies that the WER for sensitive organisms will be greater than it will be for less-sensitive ones. This is in agreement with field observations by Brungs (1991).

The WER procedure provides regulatory relief to some dischargers. However, it has 2 major deficiencies: 1) it can be expensive, and 2) it is empirical and thus cannot be used to predict future bioavailability. To have a wide spatial and temporal predictive ability, there should be a mechanistic basis for calculating site-specific criteria. Such a model should account for the effects of changes in total metal, OM, pH, hardness, alkalinity, and other chemical and physical factors on the toxicity of the metal.

Long-term solutions

Pagenkopf (1983) developed a chemical model to explain how changes in the hardness, alkalinity, and pH of the water affected metal toxicity to fish. Binding of metals to physiologically important sites on the gill results in acute mortality. The gill surfaces were treated as Lewis bases, and the cations were treated as Lewis acids. The most important point of the model is that it allows receptor sites on the gill to be modeled as ligands that react with metal ions in the water. The model predicts that increased calcium and magnesium ion concentrations, the primary components of hardness, will decrease metal uptake through competition for metal-binding sites on the gill. Indeed, trace metal toxicity decreases in the presence of higher concentrations of calcium and magnesium. Organic ligands also should reduce the toxicity of transition metals by decreasing the free-metal-ion activity. These interactions are illustrated in Figure 2-2 and can be extended to other biotic ligands or metal-receptor sites, in general. Thus, this general type of model is referred to as a Biotic Ligand Model (BLM).

The gills of freshwater fish have the 2 important physiological functions of gas transport (O_2, CO_2, NH_3) and active uptake of ions (Na^+, Ca^{2+}) (Wood 1992). Na^+ and Ca^{2+} are transported from the bulk water to the blood by active, energy-requiring "pumps." The channels or carrier proteins associated with these pumps contain specific, negatively charged regions that can complex heavy metals on the gill surface. Playle et al. (1992, 1993a, 1993b) and MacRae et al. (1999) treated the receptor sites on the gill as competitive ligands for the binding of copper and other metals. They determined conditional stability constants and site densities for these

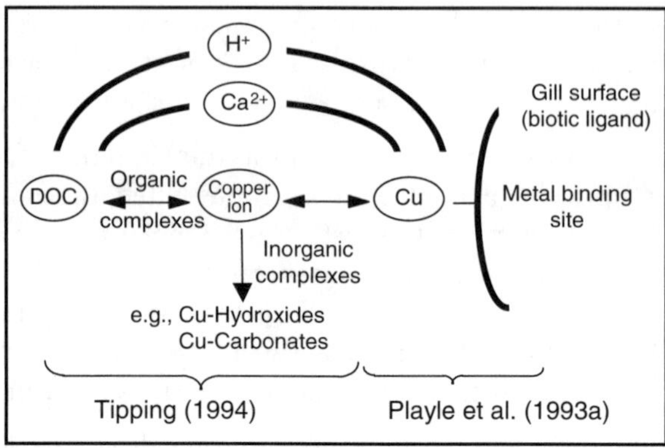

Figure 2-2 Chemical model conceptual diagram (after Pagenkopf 1983)

surface complexation reactions, and MacRae et al. (1999) demonstrated a strong relationship between the extent of saturation of the gill receptor sites and mortality.

An important conclusion from the BLM is that a specified metal species should not be treated as being bioavailable and other metal species as being not bioavailable. Rather, the presence of the gill causes chemical reequilibration in the system. It is the degree to which the gill complexation sites are occupied by metal that determines whether toxicity will occur. Although this concept is only now becoming more widely appreciated, Borgmann (1983) was the first to suggest that the fraction of receptor binding sites occupied by a toxic metal would be the same for a given biological response, independent of the water chemistry. This concept of metal-receptor interaction was a major topic at a Society of Environmental Toxicology and Chemistry (SETAC) Pellston Workshop (Reassessment of Metals Criteria for Aquatic Life Protection; Bergman and Dorward-King 1997). The participants endorsed the proposition to combine currently available geochemical speciation models (e.g., MINEQL, Schecher and McAvoy 1992; MINTEQ, Allison et al. 1991; and WHAM, Tipping 1994) and the BLM that is discussed above to provide a more comprehensive model that predicts bioavailability.

A physiologically based model that considers the bioavailability of metals in effluents and receiving waters, including the physicochemical transformations of the metal, has been developed (Di Toro et al. 2000). In this approach, metal species in receiving waters are considered to interact competitively between specific receptor sites on aquatic organisms and ligands and particulate matter in the receiving water. Thus, the model directly incorporates bioavailability concepts and explicitly accounts for variation in metal concentration and modifying factors, including pH and hardness. Meyer et al. (1999) demonstrated the validity of the underlying hypothesis that the amount of metal bound to biotic receptor sites is constant for a specified percent mortality, regardless of water quality. Chemical speciation and competitive binding of transition metals and other cations to biotic receptor sites help explain wide variations in published metal toxicity data within a given species for a specified exposure time. Analogous models will have to be developed for other classes of chemical contaminants.

There is an important need to develop models predicting body-burden and food-web transfer for compounds that have high octanol or water partition coefficients. State and federal authorities concerned with human health often regulate these compounds by identifying limits to chemical concentrations in fish tissues. To implement control strategies, however, the relationship of discharges to the level of contaminant in primary producers must be understood. This relationship must first consider the partitioning of a contaminant between the water and the organism, and second, the interaction of the hydrophobic organic contaminant with DOM. The higher the K_{ow}, the greater the affinity of the contaminant for the DOM.

This situation is even more complex for mercury for which the major concern is the concentration of methyl mercury in tissues of fish consumed by wildlife and humans. Inputs of mercury to surface waters arise from point and nonpoint sources, including the atmosphere. These inputs are predominantly metallic mercury and mercuric ion. Sediments harbor large quantities of mercury, and inorganic mercury, that can be converted to methyl mercury both in the water column and by sulfate-reducing microorganisms in anoxic sediments. Consequently, the key process required in a model to predict methyl mercury accumulation in fish is methylation. However, mechanistic models are not available to predict the rate of this process. Present models are highly site specific, requiring input of the empirically determined biomethylation rates. Analogously, rates of selenium transformations also must be empirically determined on a site-specific basis.

In summary, process-level models enable the prediction of contaminant effects despite differences in water-quality parameters among different sites, and account for bioavailability as a function of the chemical form in the water, other water constituents, and the interaction of the contaminant with the biotic receptor site. Incorporating such models into criteria will provide greater assurance that the desired level of protection in the water will be achieved. As a start, we recommend that the BLM be incorporated into WQC for metals.

Exposure Variability: Spatial and Temporal Issues

All of the approaches discussed above use steady-state models. However, pharma-codynamic or kinetic receptor-site models may be needed to account for transient and metabolic processes. Transient processes can include fluctuating exposures, such as those due to episodic discharges or spills and passage of organisms into contaminated zones for limited periods without having attained equilibrium with the contaminant. Metabolic processes include transformations of the pollutant in the environment and within the target organism.

Characterizing the spatial and temporal distribution of a chemical stressor in the environment is a necessary precursor for estimating exposure because exposure cannot occur unless chemicals co-occur with receptor species. The spatial dimension of exposure is most commonly expressed in terms of an area (or stream kilometers) that exceeds a particular toxicity threshold. However, at large scales, the spatial profile of exposure may not be appropriately expressed by area alone. The use of geographic information systems and the emerging field of landscape ecology are useful approaches that have expanded options for analyzing the spatial dimensions of exposure.

Temporal characteristics of exposure may be considered to have components of duration, frequency, and timing. "Duration" is defined as the period over which exposure exceeds some threshold concentration or over which concentration is averaged. Frequency describes how often exposure over a particular threshold occurs, either when counted in various ways as discrete events, or when measured as a percentage of time (Delos 1990). Timing of exposure, such as the order or sequence of events, also can be an important factor, particularly in relation to season and species' life cycles. For example, potentially toxic acidic conditions and monomeric aluminum concentrations have been reported to co-occur with spawning activity and presence of early life stages of striped bass in some eastern shore rivers of Chesapeake Bay, USA (Hall 1987).

In ecological exposure assessments, dimensions of intensity and time are often combined by averaging intensity over time. This produces a single number that may not adequately represent the exposure profile of a chemical in the environment and may be misleading if a receptor species is rapidly affected by a brief, high concentration of a chemical. Probabilistic frequency distributions of exposure data provide a more realistic approach for assessing potential effects on receptor species because the entire range of concentrations is considered (Solomon et al. 1996).

Spatial and temporal variability of exposure are also critical to the derivation and use of WQC. Toxicity tests used to derive criteria involve constant exposure, in contrast to the time-variable exposure likely to be encountered in the field. One method used to account for this variability is to specify a duration over which exposures are to be averaged. The management goals for aquatic ecosystems can often be attained even when these systems experience occasional mild stresses. Therefore, an allowable frequency may be specified for violations of criteria. Chapter 3 ("Time Dependence of Toxicity," p 69) discusses the importance of exposure models that deal with fluctuating concentrations and the use of time-weighted averages to assess whether criteria have been met.

Current approach for WQC development

The following discussion focuses on the approach used by USEPA to deal with the issue of exposure variability. In the U.S., WQC are currently derived from toxicity tests conducted under constant exposure conditions. The tests fall into 2 categories based on period of exposure.

Category I

Acute toxicity tests are those of 48- to 96-hour duration with survival measured as an endpoint. If tests on different life stages are available, those on the most sensitive life stages are used.

Category II

Chronic toxicity tests are those of duration over a substantial and/or sensitive portion of the organism's life cycle, measuring survival and other endpoints such as growth or reproduction. These tests usually include early life stages. In deriving criteria, effects on survival, growth, and reproduction are considered to have equal importance; other endpoints (e.g., behavioral or enzymatic endpoints) are given less consideration.

These data are used to derive 2 separate values: acute and chronic criteria concentrations. In implementation, these 2 criteria are distinguished by the duration over which ambient concentrations are to be averaged in determining attainment of the concentration: usually 1 or 24 hours for the acute criterion and 4 days for the chronic criterion.

The averaging period reflects an assumption about the relationship between toxicity and time. The 1-hour averaging period implies that a 1-hour exposure could yield the same degree of lethality as a 48- to 96-hour exposure. A contaminant's actual speed of action, as reflected by data on time to death in acute toxicity tests, is not considered in setting the averaging period. The chronic 4-day averaging period has been justified as the shortest period needed to elicit effects in chronic toxicity tests of short-lived organisms such as *Ceriodaphnia*.

The acute criterion is usually implemented as a 24-hour average when dealing with time-variable exposure at a fixed spatial location. For time-variable exposure associated with an organism moving through concentration gradients, such as mixing zones, the 1-hour averaging period is applied. An allowable frequency for violations, for example, 1 in 3 years, is also used to address the issue of time variability. In contrast to the averaging period, which reflects responses of individual organisms, the allowable frequency is intended to reflect community rates of recovery. The 3-year interval is based on the "best professional judgment" of regulatory scientists and is believed to relate to population recovery rates from catastrophic stresses.

Implementation

WQC have many uses, but it is their use in the regulatory context of setting discharge limits that is most relevant to this discussion. In the U.S., for example, WQC are used to set effluent limits enforceable in National Pollutant Discharge Elimination System (NPDES) permits and to establish cleanup goals for Superfund or other contaminated sites. This use of WQC is usually required to control discharges to receiving streams with limited flow when technology-based controls are inadequate to protect the receiving water.

The duration and frequency aspects of the WQC along with the criterion concentration (magnitude) provide the current method for taking exposure variability into account when setting effluent limits. This process is described in detail in the *Technical Support Document for Water-Quality-Based Toxics Control* (USEPA 1991). That document provides a means to translate the criterion into both daily maximum and 30-day average effluent limits. Simply stated, the method starts with the criteria duration (e.g., 1 hour or 4 days) and establishes an acceptable probability of the effluent limit being exceeded by estimating or measuring the variability of the pollutant concentration in the effluent and by knowing the frequency at which the effluent will be monitored.

This implementation process is accomplished through calculation of a total maximum daily load (TMDL). The TMDL process involves calculation of waste load allocation, allocating loads, and concentrations of contaminants among point source discharges to a given water body or watershed and a similar load allocation for nonpoint sources. Risk characterization may be applied in evaluating the modeling assumptions made in the TMDL process. These include 3 general areas: flow, mixing, and chemical transformations.

The interpretation of grab sample monitoring can be problematic because the measured violation frequency is not easily related to the criteria averaging period and frequency of exceedance. Outside of the U.S., time variability is more often handled by explicitly specifying an allowable percentage of violations, simplifying potential compliance issues but not specifically addressing the issues associated with averaging period or frequency of exceedance.

Identifying an appropriate critical flow to be used in a TMDL is a key implementation question. For point sources, low flows are usually the issue of greatest concern, and probabilistic default flows, such as the 7Q10 (7-day average low flow expected to occur once in 10 years), are often used to establish the critical flow. Because of the way in which these default flows are defined, there is a general assumption that the default flows are very conservative. This may not be the case, and flows approaching the default flow may not be as rare as the definitions imply, particularly in arid areas. For nonpoint sources, the low flow may not represent the critical event (e.g., stormwater events), and it may be necessary to look at a range of flows in the load allocation. Because default flows may not always be as conservative as generally perceived, evaluating the degree of uncertainty or conservatism applied at any particular site is a necessary component of deriving TMDLs from WQC.

Evaluation of critical flows on a seasonal basis is one way in which uncertainty or conservatism might be reduced or better described. The USEPA has developed a biologically based flow procedure that uses empirical flow data to estimate the critical flow condition. In this approach, there is no assumption about the distribution of flows. Instead, the data are reviewed to see what flows have actually occurred. Based on the averaging periods and exceedance frequency in USEPA's national criteria; the flow data are used to estimate critical flow conditions; for example, 1-day, 3-year flows and 4-day, 3-year flows. This provides a more direct match between flow and the criteria, and because there is no distribution assumption, this approach may be better suited to addressing uncertainty, particularly for rivers where flows are regulated by dams.

Mixing zones

Mixing zones are the segment of a receiving water where a discharge mixes and is diluted by the receiving stream. The spatial and temporal features of a mixing zone can be affected by many variables, including effluent and receiving water quality, flow, and temperature changes over seasons. The principal mixing zone requirements address avoidance of acutely lethal conditions, limitations on size, and restrictions on dilution assumptions. The regulation of mixing zones is usually site specific.

Mixing zones should be spatially limited to minimize adverse effects or displacement of sessile organisms. In the U.S., some jurisdictions allow an acute mixing zone—a zone of initial dilution where the acute criterion may be exceeded. The expectation is that mobile organisms, such as fish, or drifting organisms, such as daphnids, should be able to move through a mixing zone quickly enough to avoid any adverse effects. Many jurisdictions implementing WQC continue to assume that critical low flow is available when calculating permit limits. The full implications of mixing zone requirements are yet to be felt by the regulated community, and the mixing zone issue will likely become more important as jurisdictions begin to apply their requirements.

The TMDL process uses default assumptions in modeling chemical transformations of the effluent in the receiving water. These assumptions may be conservative or not, depending upon the circumstances. For example, the toxic effects of ammonia may be greatest at some distance from the point of discharge. This is because unionized ammonia is the toxic fraction, its concentration is affected by pH and temperature, and the pH and temperature in the receiving water can change significantly at some distance from the point of discharge (an increasing pH and temperature favoring more of the toxic unionized fraction). A default assumption that did not take this far-field effect into account could be underprotective.

Some jurisdictions implement their metals criteria using somewhat conservative assumptions. For example, when implementing dissolved metals criteria, they may assume that all of the particulate metal in the discharge will become biologically available after discharge. Similarly, they may assume persistence for an organic chemical that degrades relatively rapidly. When the WQC are affected by site water characteristics such as pH, DOM, hardness, temperature, etc., jurisdictions might be using conservative assumptions when deriving the values to be used for those characteristics in the TMDL calculations.

Limitations of current approach to WQC

Duration of exposure

Presently, WQC are implemented using various averaging periods to account for the temporal aspects of exposure. However, there is no specified procedure for deriving averaging periods tailored to a contaminant's speed of action. Averaging periods of 1 or 24 hours for acute criteria and 4 days for chronic criteria can be improved with existing data. Acute and chronic averaging periods probably overstate biological effects for most contaminants.

The general acute toxicity versus time model assumes that toxicity increases with the length of the exposure. Thus, with some notable exceptions such as ammonia, the longer the acute test, the lower the LC50. Exposure durations for standard acute tests with invertebrates are 48 and 96 hours with fish. This implies that for most contaminants, a 1-hour acute averaging period provides a large margin of safety. An additional level of safety is gained by dividing the final acute value by 2 to simulate an LC10 or LC1. A 1-hour exposure at the CMC[1] is not likely, therefore, to cause mortality to sensitive organisms exposed to most pollutants. Acutely toxic, rapidly acting pollutants such as ammonia can be identified and assigned shorter averaging periods.

From a practical standpoint, the 1-hour duration is difficult to implement. Effluent and ambient monitoring frequencies rarely provide real 1-hour data. The USEPA (1992) recognized this difficulty and recommended that the CMC duration be considered 1-day when acute criteria are implemented.

---

[1]CMC is the USEPA's numeric 1-hour average concentration of a compound believed to protect against acute toxic effects in exposed organisms if not exceeded more than once every 3 years (USEPA 1985).

Exceedance frequency

A specified acceptable exceedance frequency for WQC is equally critical in addressing the temporal aspects of exposure. However, the basis for the current frequency of 1 exceedance in 3 years is not adequately defined and is most likely overprotective. A significant omission in the current approach is that the magnitude of the excursion is not considered. Clearly, an exceedance 5 times the CCC will have a greater impact on the aquatic community than an exceedance of 1.05 times the CCC, all else being equal. The current approach treats all exceedances as equally harmful.

The 3-year return interval is probably too long for most WQC exceedances. The aquatic community recovery data, on which the 3-year frequency is based, are primarily from studies of catastrophic spills of toxic materials that had devastating impacts on the community (Cairns et al. 1971; Niemii et al. 1989). A minor exceedance of the CCC is very unlikely to have devastating effects on the community, and recovery is likely to take far less than 3 years. Realistic recovery periods consistent with the more typical exceedance experienced in rivers and streams need to be established. The life history patterns of receptor species also become critical when regulating the number of exceedances of a particular chemical. For example, a long-lived species with low reproductive potential, such as a sturgeon, might be impacted if a chemical threshold for juvenile or adult mortality is exceeded once in 3 years. Conversely, phytoplankton with short generation times (several days) might not be impacted by a once-in-3-year exceedance because of their rapid recovery.

Co-occurrence of sensitive life stages with exposure

The current approach for developing WQC does not consider the exposure of sensitive life stages of receptor species. This issue is crucial because the maximum likelihood for adverse effects occurs when a toxic chemical is present in the environment concurrently with sensitive life stages. For example, it would be important to know whether spawning activity and the early life stages of sensitive anadromous fish species coincide with pesticide runoff from agricultural activities. Chemical exposure not only may affect early life stages of fish but also may interfere with spawning migrations (avoidance) and recruitment. This information becomes critical for realistic regulation of a chemical, and a seasonal component might be needed for site-specific criteria.

Approaches to overcome limitations of current WQC

Near-term solutions

The following represent near-term solutions to overcome limitations of current WQC:

- Derive chemical-specific acute averaging periods using a kinetics-based toxicity model, calibrated with data from acute toxicity tests and operated on realistic exposure variability scenarios. In addition, appropriate acute mixing zones for chemicals should be derived by considering both the kinetics of toxicity and realistic exposure scenarios for organisms entering the mixing zone from the surrounding water.

- Derive an allowable frequency or an allowable percentage of criteria exceedances using data on rates of recovery from catastrophic stresses, while quantitatively accounting for the degree of population stress that realistic exposure scenarios have on taxa of various sensitivity. In absence of such a derivation, the goal may be to protect 95% of the taxa 95% of the time. However, a more stringent frequency goal might also be justified, where juvenile or adult lethality in longer-lived species is of concern in setting the criteria concentration, or where the goal is to preserve the "pristine" conditions typical of wilderness areas and national parks.

Long-term solutions

The following represent long-term solutions to overcome limitations of current WQC:

- Derive chronic averaging periods, considering kinetics and mechanisms of action for non-bioaccumulative pollutants or food chain and target organisms' uptake and depuration kinetics for bioaccumulative organic contaminants.

- Derive an allowable frequency or an allowable percentage of violations based on population modeling. Models, if validated, can be used to predict the consequences of reduced survival, growth, or reproduction for populations of taxa having various life history strategies and various degrees of toxicant sensitivity, assuming realistic exposure scenarios.

- Develop a procedure for defining chronic mixing zones for chemicals by considering the areas within concentration and effect isopleths.

Multipathway Exposure

Current approach for criteria development

WQC

Aquatic animals can be exposed to toxic chemicals in the water and in their food. The relative importance of water and food as sources depends on the relative chemical concentrations in each and the efficiency of uptake. WQC usually are derived from toxicity tests in which organisms are exposed via water only; the animals are not fed or are fed uncontaminated food. Implied by this approach is the assumption that a waterborne exposure in the lab adequately represents the environmental exposure.

SQGs

Sediment-quality guidelines have been developed in many countries to predict contaminant-related effects on benthic organisms. This goal can be achieved using principles of chemical thermodynamics and equilibrium partitioning to derive partitioning-based numerical guidelines for contaminant concentrations (Bergman and Dorward-King 1997). Guidelines exist for some organic compounds and are under active development for metals.

Sediments are biogeochemically complex, dynamic, multimedia "living" systems (Meyer et al. 1994). Chemicals are incorporated into sediments via sorption processes, biogeochemcial transformations, or organic partitioning. Concentrations of the chemicals most strongly incorporated can exceed concentrations in overlying water by as much as 10^2 to 10^6. Organisms are exposed to the concentrated pool of many chemicals in sediments by direct contact, ingestion of living or nonliving sediment components, contact with pore waters, or contact with overlying waters (Luoma 1989). The goal of the sediment guidelines has been to integrate this complexity into a single measure that accurately determines chemical exposure; however, this is an ambitious goal that may or may not be achievable.

It long has been known that the amount of chemical that an organism accumulates from sediment may not be a simple function of the total concentration in the sediment. The geochemical complexities of sediment modify the bioavailability of metals (Luoma and Bryan 1978; Di Toro et al. 1990), metalloids (Luoma et al. 1992), and organic chemicals (Di Toro et al. 1992). The nature of the contact between an organism and sediment is also highly dependent on species-specific behavior and life history characteristics (Hare et al. 1994).

The USEPA has developed SQGs that attempt to account for geochemical effects on bioavailability via equilibrium partitioning. In brief, the approach is to normalize total sediment concentrations of a chemical to concentrations of sediment components thought to most affect bioavailability (organic carbon in the case of nonionic organic chemicals; acid volatile sulfide (AVS) in the case of cationic metals). Equilibrium between the normalized sediment concentrations and porewater concentrations is assumed, and porewater concentrations are calculated on the basis of thermodynamic principles. It is further assumed that porewater concentrations are predictive for the exposure of organisms in the sediment. The calculated porewater concentrations are then compared to toxicity data derived from tests conducted during water-only exposures, to determine whether sediments exceed toxicity guidelines. Thus, sediment guidelines evaluate exposure from a multimedia system, taking advantage of simplified, but theoretically sound, geochemical principles to predict dissolved concentrations of the pollutant. However, they rely on results from single-pathway toxicity tests for the evaluation of toxicity in the system (i.e., the single-pathway toxicity test is the evaluation tool). In doing so, the guidelines assume that these waterborne exposures adequately represent the exposure experienced by animals that consume contaminated sediments.

WC

Dietary uptake is the predominant route by which piscivorous birds and mammals are exposed to bioaccumulative chemicals. This route of exposure is explicitly recognized in the procedure used by the USEPA to calculate WC values. The WC is defined as the concentration of chemical in water that, if not exceeded, protects avian and mammalian populations that use the water as a drinking or foraging source. The algorithm used to calculate the WC is an adaptation of an equation first used by the State of Wisconsin to set Wild and Domestic Animal Criteria (State of Wisconsin 1989). The Wisconsin equation was developed from methods used for human health risk assessment of compounds that act by a noncarcinogenic MOA. In its basic form, this algorithm relates a no-effect toxicant dose determined in feeding studies with surrogate laboratory species (modified by selected uncertainty factors) to estimated exposure concentrations for both drinking water and toxicant-containing food items. Feeding and drinking rates are estimated for each species and are derived from either empirical data or allometric equations. Specific BAFs are used to translate the chemical concentration in water to that in aquatic organisms for each of the trophic levels at which the test species feeds. Detailed descriptions of the procedure for calculating WC values are presented elsewhere, including guidance on the calculation and use of uncertainty factors (USEPA 1994b, 1995b, 1997). Additional documents provide exposure factors for a large number of wildlife species (USEPA 1993, 1995c).

The first use of this procedure was to calculate WC for mercury, DDT/dichlorodiphenyldichloroethylene (DDE), total PCBs, and tetrachlorodibenzo-*p*-dioxin (TCDD) in the Water-Quality Guidance for the Great Lakes System (Great Lakes Water-Quality Initiative; USEPA 1995a, 1995b). The approach was subsequently used to recommend a WC for mercury in the Mercury Study Report to Congress (USEPA 1997). A critical part of the procedure for calculating WC is the specification of BAFs for each relevant trophic level. Guidance on the estimation of BAFs for aquatic life was given in a technical support document for the Great Lakes Water-Quality Initiative (USEPA 1995d). This guidance provides for the use of laboratory data (laboratory-derived BCF × one or more food-chain multipliers [FMs]), empirical models (model-predicted BCF × one or more FMs), and field residue data. When available, field-derived BAFs are preferred to those estimated using other methods. An important advantage of using field-derived BAFs is that this confers site-specificity to the overall procedure.

Limitations of current approach to WQC

Occurrence and significance of multipathway exposure

Many early studies of contaminant bioavailability were ambiguous regarding the importance of multiple bioaccumulation pathways. Initial conceptual and quantitative models of organic contaminant bioavailability assumed that fugacity (partitioning) principles could explain concentrations attained by biota, at all trophic levels. Field studies first demonstrated that fugacity alone underpredicted bioaccumulation of some hydrophobic organic chemicals by higher-trophic-level organisms (Connolly and Pedersen 1988). It is now well recognized that trophic transfer and considerations such as food-web structure (Kidd et al. 1995) are important determinants of chemical accumulation by fishes, birds, and mammals that occupy high trophic levels in aquatic systems. This is particularly true for persistent organic chemicals and some organometallic compounds such as methyl mercury and tributyltin. Selenium is another example of a contaminant for which the importance of dietary exposure is widely recognized. The primary route of selenium uptake is dietary in birds (Heinz et al. 1987), fishes (Lemly 1985), and lower-trophic-level biota such as bivalves (Luoma et al. 1992) and copepods (Reinfelder and Fisher 1991).

The importance of dietary exposure for metals is probably the most variable of any class of chemicals. Metal uptake from food long has been demonstrated (e.g., Luoma and Jenne 1977), but the relative importance of dietary and water exposure in understanding the bioaccumulation of metals has only recently been studied. The general principles governing metals assimilation remain uncertain because they appear to be species-specific and food-specific. Moreover, dietary assimilation efficiencies of a variety of metals in feeding studies with fish tend to vary inversely with the metal concentration in food (Reinfelder et al. 1998). This observation suggests that within a range of exposure conditions, fish can regulate uptake of dietary metal. In contrast, the assimilation of organic chemicals from food is usually independent of the initial chemical concentration. Also important is the fact that the toxicity of a compound may differ depending upon the route by which it enters an animal. Compounds taken up from the diet are, by comparison to those taken up across the gills, acted upon to a greater extent by metabolic enzymes before they enter the general circulation. Obviously, this has potential to increase (sometimes) or decrease (usually) toxicity. In several studies, metals taken up across the gills and gut have been shown to accumulate in different tissues. One possible cause of these differences is route-dependent induction of metal-binding proteins. Thus, waterborne exposure to Zn appears to induce metallothionein (Bradley et al. 1985), while dietary exposures do not (Overnell et al. 1988).

Finally, aquatic organisms may be exposed to chemicals by pathways other than water or diet. For small fish, including those used in most bioassays, uptake of compounds across the skin may exceed uptake across the gills (Lien and McKim 1993; Lien et al. 1994). Dermal absorption also may contribute to accumulation of sediment-borne compounds by large fish that live in intimate contact with these sediments (Varanasi et al. 1978; Balk et al. 1984). The development of criteria for aquatic wildlife may require consideration of both dermal and inhalation (i.e., trans gill) routes of exposure. The dermal route of exposure may be particularly important for amphibians, many of which possess a thin and highly vascularized skin.

Limitations of existing criteria as a result of single-pathway assumption

The single-pathway assumption of current WQC limits its applicability to chemicals for which dietary exposure is unimportant. Significant exposure via the diet adds considerable complexity to the relationship between environmental contamination and toxicity. Although dietary exposure concentration is related to water exposure concentration, the relationship is variable, depending on the bioaccumulation characteristics of the chemical and the structure and function of the food web. Further, dietary and water exposure vary on differing temporal scales. The dietary pathway may be relatively insensitive to changes in the 4-day average concentration used to regulate for water-only exposure because of the damping of concentration fluctuations that occurs as the chemical is transferred through the food web. The extent of damping depends on toxicokinetic considerations and the number of trophic transfers between the environmental concentration (dissolved water column or pore water) and the organism of interest.

Multipathway exposures also introduce the potential for multimedia exposure. For example, bottom-feeding fish are exposed to chemicals in sediments via predation on benthic invertebrates and/or particulates and to chemicals in the water column via uptake across the gills. Changes in dietary composition with life stage also affect exposure concentration and media source. For example, lake trout fry feed on plankton, juveniles prey on water column invertebrates, and adults prey on other fish species. Maternal transfer of contaminants also is not considered.

Underestimation of delivered dose

Perhaps the most important limitation of the single-pathway approach to estimating contaminant exposure is the potential underestimation of delivered dose. Water-only toxicity tests underestimate the exposure that an animal will experience, to the extent that a chemical is bioaccumulated from dietary sources by the animal. Chemical concentrations in the food of aquatic organisms are a function of concentrations in water. Although the linkage may be variable or complex, it is possible in some circumstances to predict bioaccumulation from concentrations in a single component of the environment (water, pore water, or sediment), especially if bioavailability corrections are employed to define concentrations in the environmental media (Luoma and Bryan 1978; Tessier et al. 1993). However, this does not mean that one environmental component is the only pathway of uptake (Tessier et al. 1993). Hare and Tessier (1996) found a strong correlation between free ion cadmium and cadmium concentrations in an arthropod among several Quebec lakes. However, Hare et al. (1994) showed that dietary uptake was the primary source of cadmium in this predator, and the correlations with free ion cadmium resulted from equilibration of the various media in the lakes.

The laboratory toxicity tests typically used to derive WQC may, in some cases, represent an exposure situation that is unrealistic when compared to the "real world." In controlled exposures to a single medium, equilibration among media is not considered and cannot be reliably extrapolated. In these tests, the animal is exposed only to dissolved chemical, the dose delivered to the animal is a function of chemical concentration in water, and the criteria are derived from whatever waterborne concentration directly causes "toxicity." In nature, exposure to this same dissolved chemical dose would be accompanied by exposure to an equilibrated dietary source. The situation is more extreme as dietary exposure becomes an increasing proportion of total uptake. In fact, selenium is an example where the traditional approach to derivation of criteria had to be abandoned because of the additional dose that animals experience from diet. Standard toxicity tests of dissolved selenium indicated effects at 50 to 70 µg/L (USEPA 1992). Field studies showed that loss of diversity in a fish community and reproductive failure in surviving fish occurred at 2 to 5 µg/L (Lemly 1985). Tissue residues from the field data were employed to derive a dissolved criterion (5 µg/L) more likely to be protective of ecosystems. The extent to which the single-pathway, dissolved exposure underestimates toxicity due to multipathway exposure has not been well investigated, largely because the acceptance of dietary exposure as an important route of exposure is relatively recent.

Limitations on the derivation and use of WC

In principle, the procedure for developing WC could be applied to any predator–prey relationship if the dietary toxicity data were available and chemical concentrations in prey items could be referenced back to chemical concentrations in water. Candidates for this approach include a variety of piscivorus vertebrates and birds, bats, and amphibians that feed on emergent aquatic insects. Limitations on the derivation and use of WC include the inadequacy of the current wildlife toxicity database, the variability in BAFs at each trophic level and among aquatic systems, the natural history and dietary preference as they relate to the specification of exposure factors, and the difficulty in relating effects on individuals to effects on populations (USEPA 1994a; Abt Associates 1995). Implied by this approach is the assumption that kinetic and toxicodynamic processes in lab test species are similar to those of species in the field, although appropriate modifications are permitted if existing data suggest otherwise.

Approaches to overcome limitations of current WQC

There are 2 complementary approaches for addressing multipathway exposures: 1) derivation of criteria on a tissue-residue basis and 2) improved characterization of the toxicologically relevant dose to the animal. There is sufficient knowledge to incorporate both approaches into the criteria derivation process for selected compounds. In general, however, the tissue-residue approach is a relatively near-term solution, while the development of data and models for improved dose characterization is considered a long-term goal.

Tissue residue-based approach—a near-term solution

The tissue residue-based approach assumes that chemical concentrations in tissues can be related to a relevant toxicological endpoint. The residue criterion is then set equal to or less than this value, depending on the use of uncertainty factors. A disadvantage of this approach is that tissue concentrations and water concentrations are unspecified. In general, regulatory activities aimed at source control and reduction require knowledge of chemical concentration in water. The residue-based approach may prove extremely useful as a means of identifying systems that are out of compliance or as a check on the appropriateness of a water-based criterion for a given system. However, to do so the tissue concentration must be translated to a water concentration or tissue sampling must occur.

An important advantage of tissue residues is that they integrate variations in environmental chemical concentrations as well as natural factors (e.g., animal movements) that may complicate an exposure analysis, particularly for compounds that do not rapidly attain steady state between the organism and its environment. As such, this approach may yield much greater site-specificity than is currently attainable using exposure averaging techniques. A second advantage is that tissue concentrations of compounds that bioaccumulate are generally much higher than associated water concentrations, often permitting the measurement of chemicals that may be difficult to detect in water.

A third advantage is that no assumptions are made about the identity or bioavailability of the chemical species taken up by the animal. Because of bioavailability considerations, it is not always clear which expression of aqueous concentration is best related to toxicity. As indicated previously, high K_{ow} organic compounds bind to DOM, thus reducing their availability for uptake. Procedures have been proposed to normalize BAFs to the DOM content of water (USEPA 1995b). However, uncertainties remain due to a dependence of these procedures on empirical binding relationships, differences in DOM composition among aquatic systems, and difficulties in obtaining accurate K_{ow} estimates. The trend in aquatic toxicology research is to measure and express aqueous chemical concentrations of nonpolar organisms in terms of "freely dissolved" compound, because this is thought to best represent the concentration that is available to diffuse across biological membranes. Problems exist, however, in rigorously defining this term. In many studies, the concentration of "dissolved" compound has been operationally defined by the filtration procedures employed. Much of the compound that passes through a filter may in fact be mostly complexed with organic and inorganic ligands and may not, therefore, be "free" to diffuse. Complicating matters further, it has been suggested that bioaccumulation of some compounds is most highly correlated to the concentration of bound chemical species. For example, the extent to which selenium accumulates in aquatic biota appears to be related to factors that favor the formation and uptake of organic selenium compounds, such as selenomethionine (Lemly 1996).

Presently, the only tissue residues that are explicitly considered by USEPA in setting criteria for aquatic life and wildlife are those used to estimate BAFs in the procedure for calculating WC (i.e., BAFs based on residues in field-collected organisms). Efforts to develop WC were initiated in response to concern that existing WQC were not protective of animals that consume large quantities of fish. A common feature of the compounds for which WC have been derived is that they tend to biomagnify in aquatic food webs—fortunately, only a small number of compounds exhibit this ability. Additional biomagnification occurs between piscivorous species and their prey, often exceeding that at lower trophic levels. If, on a whole-body concentration basis, it is assumed that all animals exposed to a given compound exhibit about the same toxicological sensitivity, this biomagnification places piscivorous animals at higher risk than the animals upon which they feed.

In other situations, however, there is no a priori reason to focus attention on just one trophic level. Most compounds that bioconcentrate in aquatic biota exhibit little or no tendency to biomagnify. Additionally, the fractional contribution of the diet to total chemical uptake by an organism may vary with its trophic status, often increasing with trophic level. For example, mercury uptake directly from water probably contributes substantially to the total body burden of very small fish, while uptake by large fish is thought to be dominated by dietary sources (Post et al. 1996). Finally, differences in apparent sensitivity among species may be due to kinetic considerations (e.g., sequestration of high K_{ow} compounds in fat) or to true differences in sensitivity evaluated on the basis of chemical concentration at the site of action. In each of these instances, implementation of a residue-based approach for criteria setting would require the collection of toxicity data for a variety of taxa in a manner analogous to the present collection of toxicity data for setting WQC. Because chemical toxicity can vary with the route of exposure, it is recommended that this information be collected using protocols that reproduce the anticipated route of exposure for each of the test organisms. For a given compound, guidance on the selection of taxa also may depend on knowledge of the toxic MOA. Any effort to collect residue information for multiple species would increase the amount of testing needed to support criteria development. Presently, there is no consensus on the amount of information that would be needed for a minimum data set.

Because the calculation of a BAF requires an estimate of the chemical concentration in water, the use of BAFs to calculate WC values introduces the same problems associated with conventional WQC, including differences in chemical bioavailability and difficulties associated with measuring low chemical concentrations in water. In this respect, WC represents a hybrid of water-based and residue-based criteria-setting approaches. Similar problems will exist whenever there is a requirement to relate a chemical concentration in tissues back to the concentration of the compound in water.

A second need when using a residue-based approach is to identify the tissues to be collected and analyzed. Many factors may complicate toxicokinetic relationships among tissues, particularly when these factors lead to a chemical disequilibrium between the organism and its environment. These factors may be external (e.g., fluctuating exposure concentration) or internal (e.g., rapid growth) to the organism. The strongest correlations between residue and effect are likely to exist when chemical concentrations are measured in tissues that represent the site of toxic action. A good example of this approach is the identification of brain residues of organochlorine compounds associated with acute lethality of birds (Keith 1996). It also may be possible to sample a tissue that is a good surrogate for the target tissue. For example, blood is a good surrogate for many tissues and provides a means of characterizing chemical kinetics without the need to sacrifice the animal. This fact may be particularly important if research is to be conducted with species that are difficult to obtain or expensive to maintain. Because they are relatively easy to collect, bird eggs may provide an especially useful tissue for residue analysis. Moreover, the embryo often represents the most sensitive life stage. The interpretation of chemical residues in eggs was reviewed at a SETAC Pellston Workshop on Reproductive and Developmental Effects of Contaminants in Oviparous Vertebrates (Kleinow et al. 1999). An alternative is to sample tissues that represent sites for long-term storage or function as elimination pathways. Examples include the use of hair and feathers to characterize exposures to heavy metals. However, different species molt (or shed) with variable frequency and at different times of year, and different feather types may be replaced at different rates, even on the same individual. These observations underscore the importance of interpreting tissue residues in relation to the period of accumulation that they represent.

Dose-based approach—a long-term solution

A longer-term goal for WC derivation is to develop mechanistic models to replace the observation-based BAF concept. Additionally, there is a need to develop more exact proposals for incorporating multipathway exposures into both WQC and SQGs. All applications must incorporate the best available methods for linking contaminant concentrations in environmental media to concentrations in tissues causing adverse effects in receptor species. Presently, scientific knowledge is sufficient to incorporate qualitative (and sometimes quantitative) principles into applications of criteria, although using these principles to produce a single number for any given site is not currently realistic.

The use of exposure concentration as it correlates to toxicity requires the assumption that the rate of uptake of the chemical is the same in both the toxicity test and the field; multipathway exposure can violate this assumption. However, if absorbed dose is used as the correlate to toxicity, an uptake equality assumption is not necessary. Instead, the primary assumption is that the correlation between dose and toxicity is independent of uptake route. While this assumption may not be true in all cases, it is less strongly violated in the bioassay-to-field extrapolation than is the equality-of-uptake assumption.

In principal, dose–toxicity relationships can be developed from exposure concentration–toxicity relationships by estimating the uptake rate for the bioassay animals. The existence of substantial quantities of toxicokinetic and bioenergetic data makes such a development practical for a large number of chemicals. Thus, existing toxicity data could be used to generate a database of dose–toxicity relationships and these, in turn, could be used to derive dose-based criteria using existing WQC development procedures.

The application of dose-based criteria involves estimation of the relationship between the dose received by an organism in the field and the concentrations of the chemical in water (c_w; µg/L) and the diet (v_d; µg/g):

$$\text{dose from water (µg/g} \times \text{d)} = K_u c_w \tag{2-2},$$

$$\text{dose from diet (µg/g} \times \text{d)} = \alpha I v_d \tag{2-3},$$

where K_u is the uptake rate from water (L/g × d), α is the fraction of ingested chemical that is assimilated, and I is the ingestion rate (g/g × d). Initial values for these parameters can be estimated from laboratory experiments and bioenergetics theory (Connolly 1991). The final step in criteria application is the establishment of the relationship between the dietary concentration and the water concentration. Procedures for this step have been discussed previously in "Tissue residue-based approach" (p 36).

The dose-based approach can be extended to estimate the fate of the absorbed dose and the accumulation of injury. Such an extension has the potential to overcome the assumption that toxicity is independent of the route of uptake. Although this extension is generally beyond the current state of the science, examples exist that demonstrate its potential viability. Connolly (1985) modeled the uptake of diazinon in sheepshead minnows (*Cyprinodon variegatus*) and the reduction and recovery of brain acetylcholinesterase (AChE). Using rate constants for AChE synthesis and reaction of AChE and diazinon, the model successfully predicted the time course of AChE reduction at 3 exposure concentrations and the recovery of AChE activity after exposure ended (Figure 2-3).

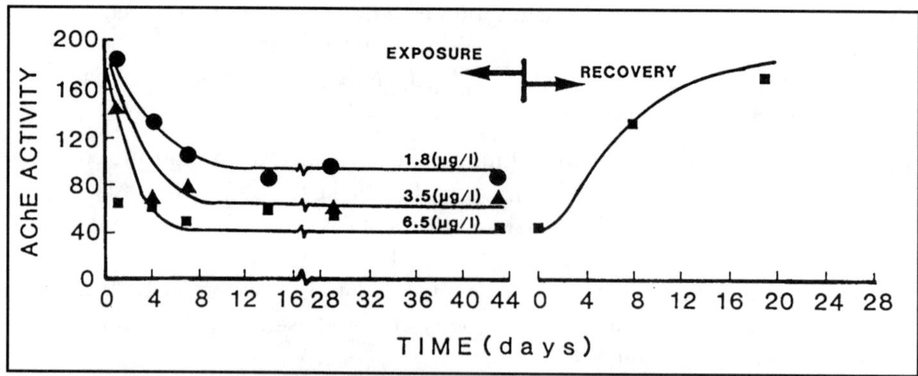

Figure 2-3 Time course of brain acetylcholinesterase (AchE) inhibition and recovery in sheepshead minnows (*Cyprinodon variegatus*) exposed to diazinon (adapted from Connolly 1985)

Physiologically based toxicokinetic (PB-TK) models hold particular promise for relating absorbed dose to the dose at the target tissue. Because these models are based on biological attributes of the exposed organism, they are well suited for extrapolating kinetic data among species. PB-TK models have been developed for a variety of common laboratory test species (Gerlowski and Jain 1983), including fish (McKim and Nichols 1994). To date, however, no attempt has been made to develop PB-TK models for birds, amphibians, reptiles, or mammalian wildlife.

Multimedia Integration

Current approach for WQC development

Currently, criteria are available for different media, but there have been few attempts to integrate across media. WQC and SQGs are an exception. The SQGs compare a dissolved porewater concentration with the dissolved concentrations predicted to be toxic in water-only bioassays. The same toxicity test data are employed in the derivation of WQC. Thus, the 2 criteria use the same information to determine toxicity, although the geochemical procedures by which they determine exposures are different (see earlier discussion of SQGs).

Recognition that chemicals are present in all media is important to criteria development for several reasons. Chemicals move among media, and environmental controls established for one medium have an impact on the concentrations in the other media. Populations and communities that the criteria are designed to protect may be directly exposed to media. Many fish are part of mixed food webs deriving energy from OM in sediments and the water column. Some wildlife species ingest a combination of benthic, pelagic, and terrestrial organisms and drink from the water column.

Intermedia transfers of contaminants occur via a multitude of advective and diffusive processes. These transfers are mediated by various physical, chemical, and biological phenomena that are highly site specific. For this reason, it is difficult to generalize the relationships among media. However, the goals of criteria development require an understanding of these relationships such that conflicts are avoided. For example, the site-specific air-quality criterion should not permit concentrations that would result in loading to a water body sufficient to cause violation of the WQC.

The focus of most criteria discussions in the U.S. is on use in the NPDES permitting process, although the actual applications may be much wider. For example, it may be useful to create integrated multimedia criteria specifically designed for managers who require forecasts or evaluations of biological effects of some future chemical discharge. Tissue residues are an important part of such evaluations, and development of these multimedia criteria would require knowledge of the relationships between residues and concentrations at the toxic receptor site in the organism, or between the residue in a prey organism and toxicity in predators, as discussed previously.

Multimedia criteria would help to frame potential discharge options for a proposed project. Criteria linked among media could be tested by evaluations for each medium followed by integration of potential outcomes into a multipathway bioaccumulation model. For example, the dissolved WQC proposed for selenium is 2 g/L as total elemental selenium. Luoma et al. (1992) suggested that particulate concentrations of selenium should not exceed 1 to 2 g/g in San Francisco Bay, USA, to avoid contamination of benthos to a level that would threaten their predators. It also has been suggested that concentrations of selenium in excess of 10 g/g dry weight in prey tissues threatens predators (Skorupa 1998). If the goal is to characterize the range of potential effects from a project with several possible selenium discharges, the range of possible loading projections from the project could be employed to estimate dissolved concentrations that might result at different input levels. These loadings then could be compared to TMDL goals, and concentrations in water could be compared to the water criterion value. Additionally, sediment–water interactive terms (K_ds) or models could be employed to derive particulate concentrations under different projected conditions. Even when these particulate chemical concentrations can be matched against criteria, they often will not satisfy concerns about a project or provide adequate detail for answers about what loads might be safe. However, particulate concentrations can be used in a multipathway model to project the range of tissue concentrations in invertebrates, and from that, one could derive an estimate of exposure to predators under different scenarios. The biological modeling brings an element directly relevant to the ecosystem into the forecasts; such projections then become more convincing to the public and more useful to managers.

The speciation of selenium in both dissolved and particulate forms affects its bioaccumulation. Speciation can be determined directly on environmental samples, but it is challenging. For the above example, assume that total selenium in dissolved and particulate forms can be projected, but that selenium speciation data were not available, or that we were trying to project some future condition for which speciation was uncertain. Multipathway bioaccumulation models could be employed to forecast tissue concentrations under different speciation scenarios based on known uptake rates for different selenium species (Luoma et al. 1992). The speciation scenario is a relevant consideration only for the first 2 trophic levels. Because dietary uptake dominates selenium bioaccumulation by predators, projections to predators can be conducted as above.

Limitations of current approach to WQC

Of particular significance is the relationship between guidelines for sediment and criteria for water. Because both are based on the application of WQC to the dissolved component of the contaminant, some level of integration exists. However, this integration does not preclude the problem of the criterion in one medium not being protective in both media. To illustrate this, consider a simple example in which the water column and sediment are represented by single, interacting compartments. Contaminant loss by degradation is ignored for simplicity. The sediment gains contaminant as the sorbed component in the water column settles into it. It loses contaminant due to resuspension and burial into deeper sediment. Dissolved contaminant exchanges between the water column and the sediment. These transfer processes are integrated in a mass–balance equation for the sediment:

$$V_2 \frac{dc_{T2}}{dt} = v_s A c_{p1} - v_d A c_{p2} + A k_m (c_{d1} - c_{d2}) \tag{2-4},$$

where the subscripts 1 and 2 designate water column and sediment, respectively; c is concentration; the subscripts t, p, and d designate total, particulate (sorbed), and dissolved components of the contaminant, respectively; v_s, v_u, and v_d are settling, resuspension, and burial velocities; k_m is the averaged effective diffusion mass transfer coefficient; V is the volume of the sediment; and A is the surface area of the sediment. Here it is assumed that the contaminant is either sorbed in an exchangeable pool on the particles or dissolved in the water (i.e., no nonexchangeable pools exist). Assuming steady state and grouping, the sediment and water-column terms, equation 2-4 becomes

$$v_s c_{p1} + k_m c_{d1} = v_u c_{p2} + v_d c_{p2} + k_m c_{d2} \tag{2-5}.$$

The particulate and dissolved concentrations are related by an equilibrium partition coefficient:

$$c_p = k_p m c_d \tag{2-6},$$

where m is solids concentration and K_p is the dry weight-based partition coefficient. Also, the settling, resuspension, and burial velocities are related by a steady-state mass balance of solids in the sediment:

$$v_s m_1 + (v_u + v_d) m_2 \tag{2-7}.$$

Substituting equations 2-5 and 2-6 into equation 2-4 and solving for the ratio of dissolved concentrations yields:

$$\frac{c_{d1}}{c_{d2}} = \frac{v_s m_1 k_{p2} + k_m}{v_s m_1 k_{p1} + k_m}$$

(2-8).

Thus, the steady-state relationship between dissolved concentrations in the water column and sediment depends on the particulate ($v_s m K_p$) and dissolved (k_m) interactions between the media. Note that chemical specificity is expressed largely through the partition coefficients. The diffusive mass-transfer coefficient is only a weak function of the chemical properties. For a contaminant with a high partition coefficient, particulate-phase transfer of chemicals between water column and sediment dominates, and the dissolved concentration ratio is equal to the ratio of partition coefficients:

$$\frac{c_{d1}}{c_{d2}} = \frac{k_{p2}}{k_{p1}}$$

(2-9).

Conversely, for a contaminant with a low partition coefficient, dissolved-phase transfer of chemical between water column and sediment dominates, the dissolved concentrations equilibrate, and the ratio of dissolved concentrations is 1.

The example indicates that for chemicals with low partition coefficients, a single-medium criterion will protect both media. For chemicals with high partition coefficients, a single-medium criterion may or may not be protective depending on the difference in partitioning between the water column and the sediment. If the partition coefficient is higher in the water column, the dissolved concentration in the water column will be lower than that in the sediment. In this case, an SQG will be protective of the water column, but a WQC will not be protective of the sediment.

The actual relationships between media depend on time-variable factors and additional processes not addressed in the above example. These factors and processes are known and can be described with varying levels of accuracy. Unfortunately, the site-specific and chemical-specific data needs are prohibitive. Simple models provide useful insights and should be used to provide a first-order indication of the linkages among media.

Approaches to overcome limitations of current WQC

Near-term solutions

Consideration of conflicts among criteria for different media is an essential part of criteria development and implementation. However, our current knowledge is not sufficient to precisely identify such conflicts. Accurate estimates of the links among media require the use of quantitative multimedia fate models. Such models exist for water–sediment systems and have been applied to individual contaminants at a number of sites. Examples include PCBs in Green Bay, Wisconsin, USA, (Bierman et al. 1992), Kepone in the James River estuary (Virginia, USA) (O'Connor et al. 1989), and zinc in the Pawtuxet River (Rhode Island, USA) (HydroQual 1995). The continued development of multimedia models is recommended to provide near-term solutions to multimedia integration.

Long-term solutions

Contaminant fate models also exist for air, and efforts to couple such models with water–sediment models are ongoing (Lun et al. 1998). However, a fully integrated quantitative multimedia fate model is not available. The use of such models requires a large investment of resources that precludes routine application in association with the application of criteria. Such application awaits the development of databases for each of the various parameters describing the modeled processes.

Research Needs

Develop and bring to proof-of-principle stage, toxicant-receptor models by chemical classes

Mechanistic models can account for bioavailability as a function of the chemical form in water, other water constituents, and the interaction of the contaminant with the biotic-receptor site. The results will enable the prediction of chemical effects despite differences in water-quality parameters among different sites. They also will provide greater assurance that the desired level of protection in the water will be achieved. Model development for some chemical classes will take longer than others. For the present, it is recommended that the BLM be incorporated into WQC for metals.

Develop a comprehensive understanding of dietary exposure

Continue to build an understanding of dietary exposure and expand the current database available for use in quantitative bioaccumulation models, with the goal of using output from such models to replace the BAF concept.

Develop a dose–effects database

Efforts must be made to convert exposure concentration-effects data to target dose-effects data. New laboratory test methods must be developed that will permit reliable determination of toxicant MOA and target dose and will provide an understanding of toxicokinetics and toxicodynamics.

Develop quantitative multimedia fate models to harmonize media-specific criteria

Existing media-specific models for water, sediment, and wildlife must be integrated by characterization of multimedia pathways. Multimedia models should be evaluated through field validation projects at well-characterized sites.

Characterize spatial and temporal scales of exposure

Monitoring in appropriate media (water, sediment, or biota, depending on chemical properties) is recommended for newly developed chemicals or chemicals lacking environmental exposure data. Validated, environmentally relevant exposure models should be used to assess the spatial and temporal scale of exposure in those instances where elevated chemical concentrations are suspected based on loadings data or source information.

Develop a comprehensive, hierarchical understanding of biological responses to contaminants

Develop the capability to relate residue–effect relationships at the tissue level to adverse impacts on organisms, populations, communities, and ecosystems.

References

Abt Associates. 1995. Technical basis for recommended ranges of uncertainty factors used in deriving wildlife criteria for the Great Lakes Water Quality Initiative. Washington DC, USA: USEPA, Office of Water. Final Report.

Adams WJ, Brix KV, Cothern KA, Tear LM, Cardwell RD, Fairbrother A, Toll JE. 1998. Assessment of selenium food chain transfer and critical exposure factors for avian wildlife species: Need for site-specific data. In: Little E, Delonay AJ, Greenberg BM, editors. Environmental toxicology and risk assessment: 7th volume. West Conshohoken PA, USA: ASTM. STP 1333.

Allen HE, Hansen DJ. 1996. The importance of trace metal speciation to water quality criteria. *Water Environ Res* 68:42-54.

Allison JD, Brown DS, Novo-Gradac KJ. 1991. MINTEQA2/PRODEFA2, A geochemical assessment model for environmental systems. Version 3.0. User's Manual. Washington DC, USA: USEPA. EPA-600-3-91-021.

Balk L, Meijer J, DePierre JW, Appelgren L-E. 1984. The uptake and distribution of [³H]benzo[a]pyrene in the northern pike (*Esox lucius*) by whole-body autoradiography and scintillation counting. *Toxicol Appl Pharmacol* 74:430-449.

Bergman HL, Dorward-King EJ, editors. 1997. Reassessment of metals criteria for aquatic life protection: Priorities for research and implementation. Pensacola FL, USA. Society of Environmental Toxicology and Chemistry (SETAC).

Bierman Jr VJ, DePinto JV, Young TC, Rodgers PW, Martin SC, Ragahunathan R, Hinz SC. 1992. Development and validation of an integrated exposure model for toxic chemicals in Green Bay, Wisconsin. Grosse Ile MI, USA: USEPA, Large Lakes Research Station. Final Report.

Borgmann U. 1983. Metal speciation and toxicity of free metal ions to aquatic biota. In: Nriagu JO, editor. Aquatic toxicology. New York NY, USA: Wiley. p 47-72.

Borgmann U, Ralph KM. 1983. Complexation and toxicity of copper and the free metal bioassay technique. *Water Res* 17:1697-1703.

Bradley R, Duquesnay W, Sprague JB. 1985. Acclimation of rainbow trout *Salmo gairdneri* Richardson to zinc: Kinetics and mechanism of enhanced tolerance induction. *J Fish Biol* 27:367-379.

Brungs WA. 1991. Synopsis of water-effect ratios for heavy metals as derived for site-specific water quality criteria. Washington DC, USA: USEPA. Contract 68-CO-0070. 40 p.

Cairns Jr J, Crossman JS, Dickson KL, Herricks EE. 1971. The recovery of damaged streams. *Association of Southeastern Biologists Bulletin* 18:79-106.

Campbell PGC. 1995. Interactions between trace metals and aquatic organisms: A critique of the free-ion activity model. In: Tessier A, Turner DR, editors. Metal speciation and bioavailability in aquatic systems. New York NY, USA: Wiley. p 45-102.

Canton SP. 1999. Acute aquatic life criteria for selenium. *Environ Toxicol Chem* 18:1425-1432.

Connolly JP. 1985. Predicting single species toxicity in natural water systems. *Environ Toxicol Chem* 4:573-582.

Connolly JP. 1991. Application of a food chain model to polychlorinated biphenyl contamination of the lobster and winter flounder food chains in New Bedford Harbor. *Environ Sci Technol* 25:760-770.

Connolly JP, Pedersen CJ. 1988. A thermodynamic-based evaluation of organic chemical accumulation in aquatic organisms. *Environ Sci Technol* 22:99-103.

Delos C. 1990. Allowable frequency for chronic criteria excursions among grab samples. Washington DC, USA: USEPA, Office of Water.

Di Toro DM, Allen HE, Bergman HL, Meyer JS, Santore RC, Paquin PR. 2000. The Biotic Ligand Model: A computational approach for assessing the ecological effects of copper and other metals in aquatic systems. New York NY, USA: International Copper Association. p 110.

Di Toro DM, Mahony JD, Hansen DJ, Scott KJ, Hicks MB, Mayr SM, Redmond MS. 1990. Toxicity of cadmium in sediments: The role of acid volatile sulfide. *Environ Toxicol Chem* 9:1487-1502.

Di Toro DM, Zarba CS, Hansen DJ, Berry WJ, Swartz RC, Cowan CE, Pavlou SP, Allen HE, Thomas NA, Paquin PR. 1992. Technical basis for establishing sediment quality criteria for nonionic organic chemicals using equilibrium partitioning. *Environ Toxicol Chem* 10:1541-1583.

Erickson RJ, Benoit DA, Mattson VR, Nelson Jr HP, Leonard EN. 1996. The effects of water chemistry on the toxicity of copper to fathead minnows. *Environ Toxicol Chem* 15:181-193.

Gerlowski LE, Jain RK. 1983. Physiologically based pharmacokinetic modeling: Principles and applications. *J Pharm Sci* 72:1103-1128.

Guy RD, Kean AR. 1980. Algae as a chemical speciation monitor–I. A comparison of algal growth and computer calculated speciation. *Water Res* 14:891-899.

Hall Jr LW. 1987. Acidication effects on larval striped bass, *Morone saxatilos*, in Chesapeake Bay tributaries: A review. *Water Air Soil Pollut* 35:87-96.

Hare L, Carignan R, Herta-Diaz MA. 1994. A field study of metal toxicity and accumulation by benthic invertebrates: Implications for the acid-volatile sulfide (AVS) model. *Limnol Oceanogr* 39:1653-1660.

Hare L, Tessier A. 1996. Predicting animal cadmium concentrations in lakes. *Nature* 380:430-432.

Heinz GH, Hoffman DJ, Krynitsky AJ, Weller DMG. 1987. Reproduction in mallards fed selenium. *Environ Toxicol Chem* 6:423-433.

HydroQual. 1995. Contaminant transport and fate modelling of the Pawtuxet River, Rhode Island. Toms River NJ, USA: Ciba Corporation. Final Report.

Keith JO. 1996. Residue analyses: How they were used to assess the hazards of contaminants to wildlife. In: Beyer WN, editor. Environmental contaminants in wildlife: Interpreting tissue concentrations. Boca Raton FL, USA: Lewis Publishers. p 1-47.

Kidd KA, Schindler DW, Muir DCG, Lockhart WL, Hesslein RH. 1995. High concentrations of toxaphene in fishes from a subarctic lake. *Science* 269:240-242.

Kleinow K, Baker J, Nichols J, Gobas F, Parkerton T, Muir D, Monteverdi G, Mastrodone P. 1999. Chemical exposure, uptake and disposition: Implications for reproduction and development in select oviparous vertebrates. In: Di Giulio R, editor. Reproductive and developmental effects of contaminants in oviparous vertebrates: Mechanisms, ecological consequences, and assessments of risk. Pensacola FL, USA. Society of Environmental Toxicology and Chemistry (SETAC). 458 p.

Landrum PF, Nihart SR, Eadie BJ, Herche LR. 1987. Reduction in bioavailability of organic contaminants to the amphipod *Pontoporeia hoyi* by dissolved organic matter in sediment interstitial waters. *Environ Toxicol Chem* 6:11-20.

Lemly AD. 1985. Toxicology of selenium in a freshwater reservoir: Implications for environmental hazard evaluation and safety. *Ecotoxicol Environ Saf* 10:314-348.

Lemly AD. 1996. Selenium in aquatic organisms. In: Beyer WN, editor. Environmental contaminants in wildlife: Interpreting tissue concentrations. Boca Raton FL, USA: Lewis Publishers. p 427-445.

Lien GJ, McKim JM. 1993. Predicting branchial and cutaneous uptake of 2,2',5,5'-tetrachlorobiphenyl in fathead minnows (*Pimephales promelas*) and Japanese medaka (*Oryzias latipes*): Rate limiting factors. *Aquat Toxicol* 27:15-32.

Lien GJ, Nichols JW, McKim JM, Gallinat JA. 1994. Modeling the accumulation of three waterborne chloroethanes in fathead minnows (*Pimephales promelas*): A physiologically based approach. *Environ Toxicol Chem* 13:1195-1205.

Lun R, Lee K, De Marco L, Nalewajko C, Mackay D. 1998. A model of the fate of polycyclic aromatic hydrocarbons in the Saguenay Fjord, Canada. *Environ Toxicol Chem* 17:333-341.

Luoma SN. 1989. Can we determine the biological availability of sediment-bound trace elements? *Hydrobiologia* 176/177:379-401.

Luoma SN, Bryan GW. 1978. Factors controlling availability of sediment-bound lead to the estuarine bivalve *Scorbicularia plana. J Mar Biol Assoc UK* 58:793-802.

Luoma SN, Jenne EA. 1977. The availability of sediment-bound cobalt, silver, and zinc to a deposit-feeding clam. In: Wildung RW, Drucker H, editors. Biological implications of metals in the environment. Springfield VA, USA: NTIS, ERDA Conference 750920. p 213-230.

Luoma SN, Johns C, Fisher NS, Steinberg NA, Oremland RS, Reinfelder J. 1992. Determination of selenium bioavailability to a benthic bivalve from particulate and solute pathways. *Environ Sci Technol* 26:485.

MacRae RK, Smith DE, Swoboda-Colberg N, Meyer JS, Bergman HL. 1999. Copper binding affinity of rainbow trout (*Oncorhynchus mykiss*) and brook trout (*Salvelinus fontinalis)* gills. *Environ Toxicol Chem* 18:1180-1189.

McKim JM, Nichols JW. 1994. Use of physiologically based toxicokinetic models in a mechanistic approach to aquatic toxicology. In: Malins DC, Ostrander GK, editors. Aquatic toxicology, molecular, biochemical and cellular perspectives. Boca Raton FL, USA: Lewis Publishers. p 469-519.

Meador JP. 1991. The interaction of pH, dissolved organic carbon, and total copper in the determination of ionic copper and toxicity. *Aquat Toxicol* 19:13-32.

Meyer JS, Davison W, Sundby B, Oris JT, Laurén DJ, Förstner U, Hong J, Crosby DG. 1994. Synopsis of discussion session: The effects of variable redox potentials, pH, and light on bioavailability in dynamic water-sediment environments. In: Hamelink JL, Landrum PF, Bergman HL, Benson WH, editors. Bioavailability: Physical, chemical and biological interactions. Boca Raton FL, USA: Lewis Publishers.

Meyer JS, Santore RC, Bobbitt JP, DeBrey LD, Boese CJ, Paquin PR, Allen HE, Bergman HL, Di Toro DM. 1999. Binding of nickel and copper to fish gills predicts toxicity when water hardness varies, but free-ion activity does not. *Environ Sci Technol* 33:913-916.

Niemii GJ, Naiman RJ, Pastor J. 1989. Factors controlling the recovery of aquatic systems from disturbance. Parts I, II, and IV. Natural Resources Research Institute. Duluth MN, USA: University of Minnesota. p 87.

O'Connor DJ, Connolly JP, Garland EJ. 1989. Mathematical models-fate, transport and food chain. In: Levin SA, Harwall MA, Kelly JR, Kimball KD, editors. Ecotoxicology: Problems and approaches. New York NY, USA: Springer-Verlag. p 221-243.

Overnell J, Fletcher TC, McIntosh R. 1988. The apparent lack of effect of supplementary dietary zinc on zinc metabolism and metallothionein concentrations in the turbot, *Scophthalmus maximum* (Linnaeus). *J Fish Biol* 33:563-570.

Pagenkopf GK. 1983. Gill surface interaction model for trace metal toxicity to fishes: Role of complexation, pH, and water hardness. *Environ Sci Technol* 17:342-347.

Playle RC, Dixon DG, Burnison K. 1993a. Copper and cadmium binding to fish gills: Estimates of metal-gill stability constants and modeling of metal accumulation. *Can J Fish Aquat Sci* 51:2678-2687.

Playle RC, Dixon DG, Burnison K. 1993b. Copper and cadmium binding to fish gills: Modification by dissolved organic carbon and synthetic ligands. *Can J Fish Aquat Sci* 51:2667-2677.

Playle RC, Gensemer RW, Dixon DG. 1992. Copper accumulation on gills of fathead minnows: Influence of water hardness, complexation and pH of the gill microenvironment. *Environ Toxicol Chem* 11:381-391.

Post JR, Vandenbos R, McQuenn DJ. 1996. Uptake rates of food-chain and waterborne mercury by fish: Field measurements, a mechanistic model and an assessment of uncertainties. *Can J Fish Aquat Sci* 53:395-407.

Prothro MG. 1993. Office of water policy and technical guidance on interpretation and implementation of aquatic life metals criteria. Washington DC, USA: USEPA. Memorandum to USEPA Water Management Division Directors, Environmental Services Division Directors, and Regions. 52 p.

Reinfelder JR, Fisher NS. 1991. The assimilation of elements ingested by marine copepods. *Science* 251:794-796.

Reinfelder JR, Fisher NS, Luoma SN, Nichols JW, Wang W-X. 1998. Trace element trophic transfer in aquatic organisms: A critique of the kinetic model approach. *Sci Tot Environ* 219:117-135.

Schecher WD, McAvoy DC. 1992. MINEQL+: A software environment for chemical equilibrium modeling. Computers. *Environ Urban Syst* 16:65.

Skorupa JP. 1998. Selenium poisoning of fish and wildlife in nature: Lessons from twelve real-world experiences. In: Frankenberger WT, Engberg RA, editors. Environmental chemistry of selenium. New York NY, USA: Marcel Dekker Inc. p 315-354.

Solomon KR, Baker DB, Richards RP, Dixon KR, Klaine SJ, La Point TW, Kendall RJ, Weisskopf CP, Giddings JM, Giesy JP, Hall LW, Williams WM. 1996. Ecological risk assessment of atrazine in North American surface waters. *Environ Toxicol Chem* 15:31-76.

State of Wisconsin. 1989. Technical support document for NR 105, 1988. Madison WI, USA: State of Wisconsin, Department of Natural Resources. Administrative Code NR 105.07.

Steemann-Nielsen E, Wium-Andersen S. 1970. Copper as poison in the sea and in freshwater. *Mar Biol* 6:93.

Stephan CE, Mount DI, Hansen DJ, Gentile JH, Chapman GA, Brungs WA. 1985. Guidelines for deriving numerical national water quality criteria for the protection of aquatic organisms and their uses. Springfield VA, USA: USEPA, NTIS. PB85-227049.

Sunda W, Guillard RRL. 1976. The relationship between cupric ion activity and the toxicity of copper to phytoplankton. *J Mar Res* 34:511-529.

Tessier AY, Couillard PGC Campbell, Auclair JC. 1993. Modeling Cd partitioning in oxic lake sediments and Cd concentrations in the freshwater bivalve *Anodonta grandis*. *Limnol Oceanogr* 38:1-17.

Tipping E. 1994. WHAM - A chemical equilibrium model and computer code for waters, sediments, and soils incorporating a discrete site/electrostatic model of ion-binding by humic substances. *Computers Geosci* 21:973-1023.

Tubbing DMJ, Admiraal W, Cleven RFMJ, Iqbal M, De Meent KV, Verweij W. 1994. The contribution of complexed copper to the metabolic inhibition of algae and bacteria in synthetic media and river water. *Water Res* 28:37-44.

[USEPA] U.S. Environmental Protection Agency. 1985. Guidelines for deriving numerical national water quality criteria for the protection of aquatic organisms and their uses. Washington DC, USA: USEPA. PB85-227049. p 98.

[USEPA] U.S. Environmental Protection Agency. 1991. Technical support document for water quality-based toxics control. Washington DC, USA: USEPA, Office of Water. EPA-505-2-90-001. 145 p plus appendices A-I.

[USEPA] U.S. Environmental Protection Agency. 1992. Interim guidance on interpretation and implementation of aquatic life criteria for metals. Washington DC, USA: USEPA.

[USEPA] U.S. Environmental Protection Agency. 1993. Wildlife exposure factors handbook. Washington DC, USA: USEPA, Office of Research and Development. EPA-600-R-93-187a.

[USEPA] U.S. Environmental Protection Agency. 1994a. Draft proceedings of the national wildlife criteria methodologies meeting; 1992 Apr 13-16; Charlottesville, VA, USA. Washington DC, USA: USEPA, Office of Water and Office of Science and Technology. 150 p.

[USEPA] U.S. Environmental protection Agency. 1994b. Interim guidance on determination and use of water-effect ratios for metals. Washington DC, USA: USEPA, Office of Water. EPA-823-B-94-001.

[USEPA] U.S. Environmental Protection Agency. 1995a. Final water quality guidance for the Great Lakes system: Final rule. *Federal Register* 60(56):15366-15425.

[USEPA] U.S. Environmental Protection Agency. 1995b. Great Lakes Water Quality Initiative criteria documents for the protection of wildlife: DDT; Mercury; 2,3,7,8-TCDD; PCBs. Washington DC, USA: USEPA, Office of Water and Office of Science and Technology. EPA-820-B-95-008.

[USEPA] U.S. Environmental Protection Agency. 1995c. Great Lakes Water Quality Initiative technical support document for wildlife criteria. Washington DC, USA: USEPA, Office of Water and Office of Science and Technology. EPA-820-B-95-009.

[USEPA] U.S. Environmental Protection Agency. 1995d. Water quality guidance for the Great Lakes system: Supplementary information document (SID). Washington DC, USA: USEPA, Office of Water and Office of Science and Technology. EPA-820-B-95-001.

[USEPA] U.S. Environmental Protection Agency. 1996. Proposed selenium criterion maximum concentration for the water quality guidance for the Great Lakes Systems; proposed rule. Washington DC, USA: USEPA. 40 CFR Part 132, 61 (221):58444-58449.

[USEPA] U.S. Environmental Protection Agency. 1997. Mercury study report to Congress. Washington DC, USA: USEPA, Office of Air Quality Planning and Standards, and Office of Research and Development.

Varanasi U, Uhler M, Stranahan SI. 1978. Uptake and release of naphthalene and its metabolites in skin and epidermal mucus of salmonids. *Toxicol Appl Pharmacol* 44:277-289.

Wood CM. 1992. Flux measurements as indices of H^+ and metal effects on freshwater fish. *Aquat Toxicol* 22:239-264.

CHAPTER 3

Effects Assessment

David R. Mount, Gerald T. Ankley, Kevin V. Brix, William H. Clements,
D. George Dixon, Anne Fairbrother, Christopher W. Hickey, Roman P. Lanno,
Christopher M. Lee, Wayne R. Munns, Robert K. Ringer, Jane P. Staveley,
Christopher M. Wood, Russell J. Erickson, Peter V. Hodson

Introduction

The Effects Workgroup reviewed the state of the science with regard to the measurement and prediction of ecological effects, in the context of WQC development. The purpose of this review is to identify where substantive areas of uncertainty exist and to indicate what information could be applied to reduce these uncertainties in the immediate term, near term, and far term.

The consensus of the workgroup was that the general approach of establishing WQC on the basis of laboratory toxicity data is a reasonable approach for continued use. However, the group cited many specific instances of how and when this approach may not provide the desired level of ecological protection. Several enhancements were identified that could reduce uncertainty and improve the ability of WQC to protect natural assemblages without introducing unnecessary conservatism. The scientific basis for some changes is available for incorporation today or will become available in the near future, while others will require substantial additional research and may take 5 to 10 years for development and incorporation. The general focus of all changes is to increase accuracy, decrease uncertainty, and express the residual uncertainties as explicitly as possible.

Reevaluation of the State of the Science for Water-Quality Criteria Development. Mary C. Reiley et al., editors.
©2003 Society of Environmental Toxicology and Chemistry (SETAC). ISBN 1-880611-30-9

The consensus of the work group can be broadly summarized in 4 primary recommendations:

1) Information on the expected environmental behavior and toxic mechanisms for prospective criteria chemicals should be used to influence the design of criteria and the strategy for data collection. Convening an expert panel at the beginning of WQC development may facilitate this process.

2) To enhance the analysis and interpretation of toxicity data, the WQC development process should be amended to a) use regression analysis instead of hypothesis testing for analysis of chronic toxicity data, b) incorporate time-response data, c) express statistical uncertainty explicitly, d) require replication of critical studies, and e) consider residue-based criteria as an approach for certain chemicals.

3) Bioavailability, as reflected in water chemistry or tissue residues, should be incorporated into WQC to maximize their site specificity.

4) To increase ecological relevance, data collection and criteria derivation should be expanded to a) include additional taxa, especially plants and amphibians; b) include effects on wildlife species, especially from bioaccumulative chemicals; c) consider mixtures of chemicals instead of just single chemicals; d) incorporate evaluation of dietary pathways and nutrition; and e) develop approaches for integrating endpoints beyond survival, growth, and reproduction for all species, using tools such as microcosms, mesocosms, field studies, and physiological measures.

Topical discussions within this chapter parallel the structure of these recommendations. Specific recommendations and research needs identified by the group are given in bullet form at the end of the applicable section.

Determining Approach and Data Needs for Effects Assessment

Overview

For most jurisdictions, the derivation of WQC requires a minimum data set. For example, the USEPA requires at least 8 acute and 3 chronic tests with aquatic species representative of a broad phylogenetic range (Stephan et al. 1985). While this approach seems reasonable for chemicals for which little is known about physicochemical properties or toxic MOA, it is somewhat naive for chemicals for which there are existing data or for which predictions of physicochemical properties or MOA can be made. Selection of test species and endpoints based on this type of information can optimize resources and increase the accuracy of criteria. This requires, however, that existing data and models for chemicals of concern be considered early in the criteria derivation process, rather than after the completion of testing.

Bioaccumulation potential

Physicochemical properties related to the expected distribution and persistence of a chemical in aquatic ecosystems must be considered before deciding whether a criterion is even required. These properties also are critical in the design of tests to assess potential toxicity to invertebrates, fishes, and wildlife. A key variable is the propensity of a chemical to bioconcentrate or bioaccumulate, which influences the type of criterion that would be most appropriate (e.g., water-based versus residue-based). Highly bioaccumulative chemicals are generally of more concern at higher trophic levels than at low, so data collection and criteria derivation need to emphasize assessment of effects at higher trophic levels. For example, potential effects on wildlife species should be evaluated routinely for more highly bioaccumulative chemicals (see "Criteria for the Protection of Wildlife," p 91).

The potential for bioaccumulation can be assessed either from empirical data or through predictive models. There are bioconcentration or bioaccumulation data for a wide variety of chemicals in the literature, collected from both laboratory and field studies. For chemicals that are not easily metabolized or excreted, there are relatively robust models that estimate bioconcentration potential based upon the octanol–water partition coefficient (K_{ow}) of a chemical (e.g., Veith and Kosian 1982). While assessment of chemicals that are readily metabolized and excreted is more complicated, there are models being developed that can help evaluate the potential for metabolism (e.g., Schüürmann and Schindler 1993; De Groot et al. 1995; Hollebone et al. 1995; Mekenyan et al. 1995).

Properties that make certain chemicals likely to bioaccumulate also make them likely to sorb to sediments. Accumulation of chemicals by pelagic species will be influenced by the degree of partitioning from the sediment and by transfer into the pelagic food chain from benthic organisms. Both processes are clearly influenced by concentrations of chemicals in both water and sediment. For this reason, criteria for most highly bioaccumulative chemicals will require integration of WQC with criteria for sediments. The concern for this chapter is selecting appropriate species and methods for evaluating effects. More details regarding estimating bioaccumulation, bioavailability, and multiple routes of exposure are found in Chapter 2, Exposure Analysis.

Residue-based criteria

Residue-based criteria have been proposed as an alternative to water concentration-based assessment for hydrophobic organic and organometallic chemicals, for which exposures are often significant from both water and diet (McCarty and Mackay 1993). Residue criteria are based on the concept of CBRs, the concentrations of toxicants in organisms that result in a specific toxicity endpoint such as lethality. CBRs explicitly consider bioavailability and integrate exposure to toxicants over time rather than representing point estimates of exposure after specific periods. For some hydrophobic chemicals (e.g., PCBs), the development of water-column-based criteria is intuitively incorrect, since their uptake is influenced by concentrations in sediment and diet as well as in water. CBRs may offer an alternative to WQC for these compounds and could be used to identify unacceptable exposures for aquatic organisms. Many jurisdictions already limit the concentration of some compounds in tissues of prey species to protect both wildlife and human consumption (e.g., the International Joint Commission).

The CBR approach has definite advantages when attempting to determine the bioavailability and toxicity of chemicals for which the environmental and pharmacokinetic parameters that control uptake are not fully understood; adequacy of environmental protection can be assessed directly by measuring tissue residues in free-living organisms. The CBR approach is applicable when the toxic effect of a particular residue is the same regardless of the route of uptake. Chemicals that act by narcosis (e.g., organic solvents) are thought to contribute to the same tissue compartment (lipoprotein membranes) regardless of the route of uptake (Van Wezel and Opperhuizen 1995). This may not be true for some chemicals such as copper or selenium (Hodson and Hilton 1983; Miller et al. 1993) where dietary and waterborne routes of uptake influence the modes of action and toxicities.

Critical body residues can be estimated as a simple extension of both acute and chronic toxicity tests. Organisms should be preserved for tissue residue analysis during and at the completion of toxicity tests. Residue analysis can complement both acute and chronic dose–response data (Figure 3-1). Toxicokinetic modeling can be applied to CBR data, providing a method for the development and testing of hypotheses, particularly for defining MOA and studying the toxicity of mixtures. Compilations of residue and effect data are available to support this effort (e.g., Jarvinen and Ankley 1999).

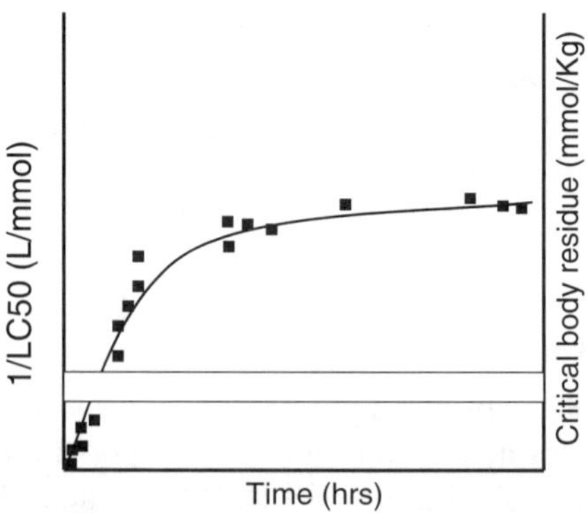

Figure 3-1 LC50 varies with exposure time (data points and curve), while CBR (horizontal line) is independent of time.

By themselves, CBRs may provide an excellent means of assessing compliance with WQC by demonstrating whether accumulated residues in native organisms are at levels associated with adverse effects. However, to apply CBR-based WQC as standards in effluent discharge permits, remedial actions, or similar applications, it will be necessary to demonstrate the association of tissue concentrations with chemical concentrations in various environmental media (see Chapter 2). This association may be confounded by interspecific differences in equilibration times (e.g., Connell 1988).

Toxicological considerations—MOA

Several clear examples from the literature and previous criteria show that MOA information could have guided the development of WQC. One current example concerns chemicals that interact with the estrogen receptor (ER) to produce a host of physiological alterations, including potentially adverse reproductive effects in fish and wildlife. Because many current WQC testing guidelines do not require a vertebrate (fish or wildlife) reproductive assay as part of the standard suite of tests, criteria so derived may not protect against ER-mediated effects. Comprehensive assessment of these chemicals requires testing with full life-cycle (reproductive) tests on vertebrates, or with suitable screening assays sensitive to these effects.

Another example in which MOA information could guide species and endpoint selection is Ah receptor (AhR) agonists, which include a variety of polychlorinated dibenzo-*p*-dioxins (PCDDs), polychlorinated dibenzofurans (PCDFs), PCBs, and other structurally similar compounds. AhR agonists are extremely toxic to early life stages of vertebrate species and can induce delayed mortality that would not be observed effectively during a 96-hour assay. Based upon this knowledge of MOA, however, it is possible to develop a suite of tests with appropriate sensitivity. For example, effort would be wasted on extensive testing of AhR agonists with invertebrate species, which lack the AhR and, therefore, would be insensitive (Cook et al. 1993; West et al. 1997). Further, because adult fish are at least 10-fold less sensitive than early life stages (Cook et al. 1993), assays that encompass early life stages are essential. Finally, because of delayed mortality, standard 96-hour lethality tests are insufficient and should be extended when evaluating AhR agonists (e.g., Elonen et al. 1998).

Other chemical classes for which existing toxicological data should drive species and endpoint selection include compounds that are designed to exploit specific MOAs, such as organophosphate insecticides, invertebrate growth regulators, and herbicides. Existing toxicological data show that crustaceans, particularly cladocerans, are very sensitive to organophosphate pesticides and other acetylcholinesterase inhibitors (e.g., Ankley et al. 1991; Amato et al. 1992). Insect growth regulators, developed as pesticides, should receive particular attention for their possible effects on nontarget invertebrates, which have the same physiology as the target species (Wigglesworth 1970; Bowers 1990). Finally, herbicides, which are designed to exploit plant physiology, should be assessed using tests with aquatic plants; unfortunately, there are few standardized test procedures for aquatic plants. Algal assays alone may not provide an adequate substitute for tests with higher plants.

Better selection of appropriate test species and endpoints is a step that must be incorporated into the earliest stages of criteria testing; current approaches often are based solely on existing data from standardized tests, and hence other relevant information may not be considered until after testing has been completed. Relevant information could consist of toxicity data from nonstandard tests (e.g., those with nonindigenous species) or data concerning endpoints other than survival, growth, and reproduction (e.g., induction or inhibition of specific enzyme systems). In most cases, this type of information is not used for criteria derivation but could be extremely helpful when designing a program of toxicity testing. For example, subcellular responses (sometimes referred to as "biomarkers") can suggest a specific toxic MOA. Induction of certain classes of cytochrome P450, for example, is typical of AhR agonists (see Cook et al. 1993 for review); as described above, this could direct the selection of appropriate species and endpoints. Similarly, induction of vitellogenin (egg yolk protein) indicates exposure to ER agonists in egg-laying vertebrates (e.g., Herman and Kincaid 1988; Donohoe and Curtis 1996; Nimrod and Benson 1998) and the need for vertebrate reproductive assays in the suite of tests used for WQC derivation. Identification of possible MOA need not be limited to information from tests with aquatic or wildlife species. For many chemicals, this type of information could be acquired from mammalian tests; the commonality in toxic MOA across vertebrates serves as a basis for across-species extrapolation, especially during the problem formulation phase of assessments (Ankley, Johnson et al. 1997).

For chemicals for which no toxicity data exist, some insight as to MOA could be gained through structure-activity relationship (SAR) models or in vitro assays that focus on a specific endpoint. For example, there are several SAR models that identify whether the structural characteristics of a chemical would be consistent with binding to, and activation of, a number of receptor-based systems associated with adverse effects (for review see Ankley, Bradbury et al. 1997). Similarly, SAR models can predict whether chemicals should exert toxicity through narcosis, reactive mechanisms, or photo-activated toxicity (Bradbury 1994, 1995; Mekenyan et al. 1995). In addition to mechanism-specific SAR models, there are also a variety of inexpensive in vitro assays at different biological levels of organization (e.g., receptor binding, gene transcription, translation, mutagenicity) that could assist identification of potential toxic MOAs (e.g., see review by Zacharewski 1997). Of course, the use of such tests also would require a deviation from the existing WQC testing approach, involving the collection of new data before actual testing for criteria derivation commenced. Ultimately, though, this could prove very cost-effective in the context of better defining appropriate types of in vivo assays.

As a general note of caution, the design of a test suite for criteria derivation should be flexible enough to enable consideration of "unexpected" responses that might not be consistent with predicted or presumed MOAs. For example, recent studies with the herbicide atrazine suggest that it acts as an endocrine disrupter in mammalian systems (Cooper et al. 2000). Hence, in addition to survival and growth tests with plants, fishes, and perhaps wildlife, reproductive tests are needed before deriving a criterion for atrazine. In another example, it appears that the insect growth regulator methoprene also might affect vertebrate developmental pathways, in particular those under control of retinoids or retinoic acid receptors (Ankley and Giesy 1998). These examples highlight the need to maintain some level of testing across a broad range of phylogenetic groups, even for chemicals with seemingly specific MOAs.

Another type of existing data that could be used in the initial design of tests for criteria derivation is that derived from mesocosms, microcosms, and field studies; these approaches are currently little used in most WQC development schemes. As described in "Population, Community, and Ecosystem Responses" (p 101), the increased scope of responses addressed by mesocosms and other ecologically based studies can be employed to assess both direct and indirect effects induced by chemical exposure.

Role of an expert panel

Expanding the use of existing data and/or generation of new data for the design of tests for WQC derivation requires that in-depth professional judgment be used in all stages of the assessment, not just in the final analysis. It is recommended that this judgment be provided by an expert panel of scientists charged with addressing the issue of appropriate evaluation steps for specific chemicals of concern before testing commences. This expert panel should consist of scientists knowledgeable about the chemical or class of chemicals under consideration and should represent a cross-section of interested parties. The panel should be involved in test design, test interpretation, and final review of the derived criteria. This would ensure consideration of other existing data for the chemical of concern, enable a significant degree of up-front technical input, and provide a level of peer review that should facilitate wider and more ready acceptance of the recommended criteria.

Recommendations

- An expert panel should be convened prior to the development of a criterion; this panel should play an integral role in defining appropriate evaluation steps for criteria derivation.

- The approach to criteria derivation must recognize the expected environmental behavior of the chemical and the potential need for integrating WQC with companion criteria (e.g., sediment, wildlife).

- The suite of toxicity tests used for criteria derivation should not rely solely upon generic guidelines (e.g., Stephan et al. 1985) but should be tailored to capture MOAs or endpoints affected by the chemical of concern.

- Careful thought should be given to the types of chronic toxicity data required for criteria derivation, to insure that appropriate critical life stages or processes are considered.

Measuring and Interpreting Chronic Toxicity

Although many approaches to WQC derivation include assessment of both short-term ("acute") and long-term ("chronic") effects, all recognize that sublethal effects of chemicals occurring over long exposure periods represent a threat to aquatic communities, and that WQC must protect against those effects. However, compared to acute toxicity data, chronic toxicity data are more resource-intensive to collect, more varied in their endpoints, and may be more difficult to interpret in terms of ecological significance. Moreover, there are many species for which chronic test procedures have not been developed. From this, it should be no surprise that there is substantial uncertainty in how best to design, collect, and interpret chronic toxicity information in the context of WQC.

Uncertainty regarding chronic toxicity can be broadly classified into 2 areas: 1) uncertainty pertaining to the way in which chronic toxicity is quantified and expressed, and 2) uncertainty regarding how to incorporate chronic toxicity measurements into the criteria derivation process. Both of these issues may be influenced by the relationship between chronic toxicity endpoints and ecological responses.

Quantification of chronic toxicity

Most chronic toxicity data are evaluated using hypothesis testing to derive the NOEC and the LOEC from the test concentration series. These estimates are dependent in large part on the design of the tests themselves, with the number and progression of concentration treatments, the degree of sample replication, etc., influencing the power of the test to detect differences among treatments. Tests with low variability may result in a LOEC representing responses that are only 2% to 3% different from the control, while tests with high variability may result in an NOEC corresponding to a response greater than 40% different from the control. Thus, the NOEC is not the highest chemical concentration causing no observable toxicological effect, but rather is the highest treatment within the chronic test that is not statistically different from the control response. Probability of Type II error (false negative) is generally not calculated or reported. Calculated NOECs therefore may not be reasonable estimates of the threshold concentration below which no effect would be expected. For example, fish reproduction data are notoriously variable, yet reduced reproduction may be one of the most ecologically significant toxicological responses, at least for certain species (Suter et al. 1987). Hence, the level of protection afforded by NOEC or LOEC data may be driven by study design and data precision rather than by ecological relevance or policy decision. This issue has been recognized in the human health arena for some time and is receiving considerable attention in aquatic toxicology (Stephan and Rogers 1985).

A more rigorous approach to estimating threshold response involves development of full exposure–response relationships using the entire chronic test data set. Based on methods similar to those employed for point estimation of acute lethality, regression models can be used for sublethal responses. Levels of response (e.g., effective concentrations EC5, EC20) can be estimated from these models with associated uncertainty bounds (Figure 3-2A).

The use of regression analysis and point estimation, as an alternative to hypothesis testing, was discussed extensively and was viewed favorably in a previous workshop on effluent toxicity testing (Grothe et al. 1996), and the principles and issues applicable to effluents are applicable to chemical-specific testing as well. Moore and Caux (1997) assessed how well a variety of different models for sigmoidal exposure–response curves fit a series of data sets developed for new product notification or published in the peer-reviewed literature. They concluded that regression analysis was superior to hypothesis testing for deriving realistic thresholds with well-defined variance to effect levels as low as 5% to 10%; below 5%, the variance of estimates increased sharply, particularly where data were extrapolated rather than interpolated. The analysis also demonstrated that experimental designs for hypothesis testing,

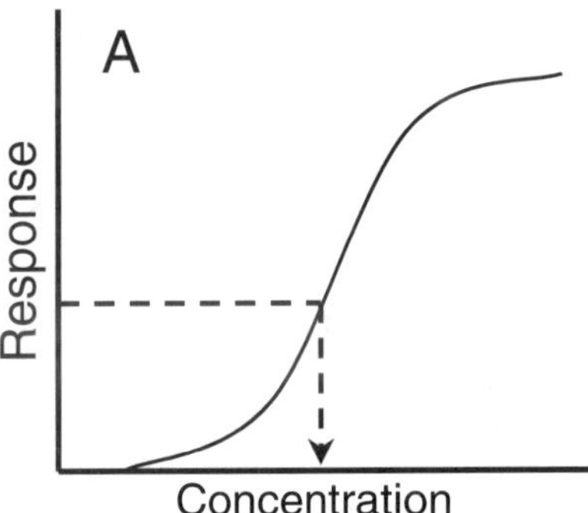

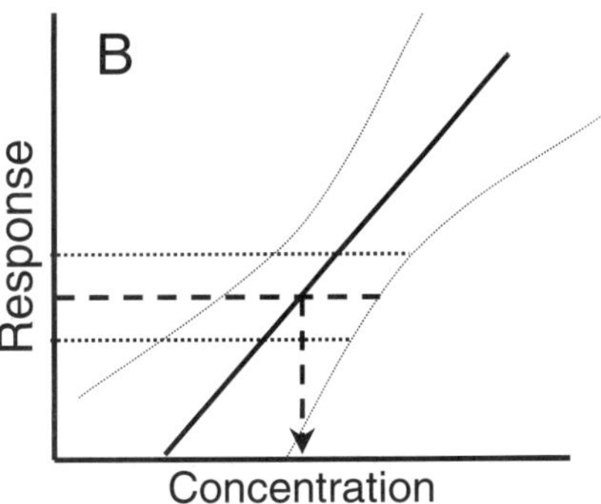

Figure 3-2 Possible approaches for developing chronic threshold concentrations–A) continuous exposure–response model, B) 2-phase model. Broken arrows indicate possible chronic thresholds.

which emphasize multiple replicates, limited numbers of treatments, and widely spaced treatments, were inappropriate for regression methods. Regression models work best when treatments are closely spaced and cover a broad range. Therefore, to improve future data for setting criteria, new experimental designs and standard protocols are required. While more exposure levels will be tested in a regression model, the total experimental effort would likely remain the same because multiple replicates (usually 3 to 5) can be replaced by multiple exposure levels.

Several advantages accrue from the development of chronic exposure–response relationships. In addition to reducing uncertainties associated with estimates of threshold concentration (as well as enhancing definition of those uncertainties), these relationships use all of the information available from a chronic test. Further, knowledge of exposure–response relationships permits estimation of the chemical concentration causing any specified level of effect. Different levels of biological effect may be appropriate for different assessments. In addition, the degree of anticipated effect can be estimated in cases where criteria are exceeded.

Another alternative for describing chronic threshold concentrations is to select the effect threshold based on the variance associated with a specific test. For example, if for a given test (or test procedure) the minimum significant difference (MSD) is estimated to be 17%, then an estimate of the EC17 could be used as the chronic threshold for that test. This approach would be an improvement over the NOEC or LOEC approach because it reduces the importance of test concentration selection by providing a continuous, rather than a discrete, response variable. However, like the NOEC or LOEC, it is still heavily influenced by the degree of replication and the inherent variability of the response variable and does not recognize the ecological consequences of the reported effect. MSD analysis also could be used to set bounds on acceptable variability within a test. For instance, tests in which the MSD was less than 5% or greater than 35% might be considered unacceptable for inclusion in WQC data sets.

Other approaches for establishing threshold (or other) response also may be valuable. For instance, general linear model estimation procedures could be used to establish the threshold "breakpoint" of a 2-phase model (Figure 3-2B) (e.g., Bennett and Schafer 1988). Caution is needed in development and application of these procedures to ensure that errors in model selection are minimized; this represents an important research need with respect to estimation of chronic thresholds. Valuable insight into these issues might be gathered from developments in other fields (e.g., pharmacology).

Incorporation of chronic toxicity into WQC derivation

From a conceptual standpoint, there are 2 basic approaches to establishing WQC to protect against chronic effects. In the first approach, threshold effect concentrations are divided by numerical factors to account for specific uncertainties (e.g., acute to chronic extrapolation, relative species sensitivity). These may be fixed values (e.g., 10) or may be calculated from available data (e.g., the ACR approach described below [Stephan et al. 1985]). The second approach involves statistical analysis of the distribution of available data to project a concentration that will be protective at a specified level. In general, numerical adjustment approaches are considered to be less rigorous technically, but may be necessary in cases where insufficient chronic data are available to use statistical techniques, a common occurrence given the cost and technical constraints involved in generating chronic data for large numbers of species. For example, under USEPA guidelines, only 2 WQC have been derived directly from chronic data (without the use of ACRs), and even in those cases, data were not available for all 8 of the required taxa.

Several procedures for estimating levels of protection from the distribution of chronic toxicity endpoints have been described. In broad terms, chronic data from all tested species are fit to a statistical distribution; then, an extrapolation or interpolation approach is used to estimate a given percentile within the sensitivity distribution (e.g., Figure 3-3; Aldenberg and Slob 1993; Organization for Economic Cooperation and Development [OECD] 1995). Criteria derived from statistical analysis of large data sets of chronic toxicity should provide the most accurate criteria with the least uncertainty. In addition, many of the statistical techniques can be used to generate uncertainty bounds around the point estimate for the distribution (see Figure 3-3 and Chapter 4, Risk Characterization).

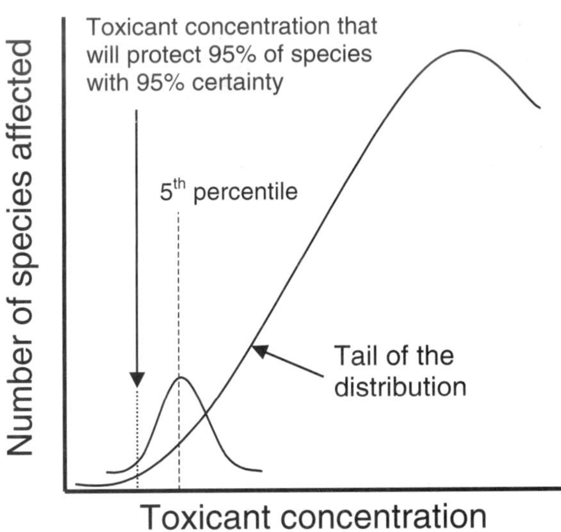

Figure 3-3 Example calculation of criterion concentration from distribution of chronic data. Smaller bell curve represents the uncertainty around the 5th percentile of the larger curve. (After Aldenberg and Slob 1993; OECD 1995)

Error associated with the use of the statistical regression-fitting models (e.g., Aldenberg and Slob 1993) comes from 2 major sources:

1) the goodness of fit of the log-normal or log-logistic model to the data (this problem can be overcome by testing whether the data are likely to fit a logistic distribution and by visual inspection of the data); and

2) the leveraging of the fitted relationship. Because the logistic distribution is symmetrical, both high and low outliers may bias the relationship. Again, it is important to check the distribution of the data visually.

In application, these statistical regression models provide an uncertainty calculation with optional confidence intervals around the specified level of protection (e.g., 95% of test species). The size of this uncertainty interval is dependent on the number of data, their variability and goodness of fit to the distribution model. In contrast, the application factor (AF) approach (Canadian Council of Resource and Environment Ministers [CCREM] 1987) and the triangular distribution (Stephan et al. 1985) provide no indication of uncertainty relating to the number of data incorporated in the calculation procedure.

There are often insufficient chronic toxicity data available to calculate WQC directly. Under USEPA WQC guidelines, the ACR is used to estimate chronic thresholds. The ACR is the ratio between the acute LC50 and chronic value (ChV; geometric mean of the NOEC and LOEC) for a given species in which both endpoints were established in the same laboratory under similar test conditions. To generate a WQC using the ACR approach, USEPA guidelines require ACRs for at least 3 families, including at least 1 vertebrate, 1 invertebrate, and a third species with high sensitivity to acute exposure. A final ACR is derived from these values and applied to the final acute value (FAV) to develop a final chronic value (FCV).

The ACR approach has been criticized as lacking scientific rigor. One of the primary concerns is that applying a single ACR to an acute toxicity threshold derived from a variety of species assumes that the MOAs for acute and chronic toxicity are common across species. In reality, this is probably not the case for many chemicals (see previous discussion in "Toxicological considerations— MOA," p 57). Consequently, if chronic toxicity tests were performed on all of the species represented in the acute toxicity data set, a different species distribution and subsequent WQC would likely result. There are other difficulties that may arise in selecting acute and chronic toxicity data for developing the ACR. For example, acute toxicity tests may be conducted with varying life stages (e.g., larval, juvenile, adult) while chronic toxicity tests are based on differing endpoints (e.g., growth and reproduction). Different combinations of these life stages and endpoints will lead to variability in calculated ACRs. For these reasons, WQC derived using ACRs will have an inherent uncertainty that is not easily quantified or expressed in the final criterion.

Some of these weaknesses are recognized in the USEPA approach, and the guidelines discuss options to reduce potential biases. Derivation of the final ACR involves more than simple averaging and is dependent on the distribution of individual ACRs. For some chemicals, acutely insensitive species have considerably larger ACRs than acutely sensitive species (e.g., copper). When this is the case, only ACRs for the acutely sensitive species are used to develop the final ACR. Sometimes ACRs follow taxonomic patterns (e.g., mercury), such as fish showing consistently higher ACRs than invertebrates. In these cases, ACRs for different taxa can be applied to acute data for the same taxa to determine an appropriate criterion. Where ACRs vary by more than a factor of 10 across species and there are no clear associations with acute sensitivity or taxonomic groups, the guidelines state that a criterion cannot be derived using the ACR approach.

While the ACR approach has identifiable weaknesses, it may offer at least one advantage over direct analysis of chronic data. Where the WQC derivation proceeds directly from chronic data only, the data set is limited to only those species for which chronic test methods are available. Hence, if there are highly sensitive species that for logistical or practical reasons are not tested for chronic toxicity, their sensitivity will not be reflected in the final criterion. When an ACR approach is used, data from organisms with high acute sensitivity but unknown chronic sensitivity can be reflected in the final criterion. For example, in the USEPA criteria document for copper (USEPA 1984), only 3 of the 10 most acutely sensitive genera have chronic data available, and 5 of the 10 genera are organisms for which chronic methods are not available. Overall, acute data for copper are available for 52 species covering 41 genera, while chronic data are available for only 9 species covering 8 genera. There are no chronic test methods for several freshwater mollusks, insects, amphibians, and a wide variety of marine organisms, and as such, these species will not be represented in analyses based on chronic toxicity alone, unless some form of extrapolation is used.

From a scientific perspective, the most rigorous criteria will result when chronic toxicity data are available for a broad range of organisms; development of the methodologies and data to support this approach should be a priority. At present, however, it is difficult to meet even the 8-family requirements for chronic toxicity data outlined by the USEPA (Stephan et al. 1985), particularly for marine organisms.

For this reason, correction of the chronic data set limitations will be a long-term goal. In the short-term, one solution to the issue of limited data would be to lower the minimum data requirements for deriving chronic criteria. This solution needs careful consideration because limited data sets will result in greater uncertainty and may lower chronic criteria simply as a function of these data limitations. Under most statistical approaches used for WQC derivation, the criterion resulting from databases with only 8 to 15 species will be considerably lower than the lowest ChV in the data set. Chronic criteria may be lowered further because of increased statistical uncertainty (low numbers of species) even when acute toxicity data are abundant. Perhaps more importantly, reducing the phylogenetic coverage of the minimum database increases the probability that important, sensitive taxa may be missed. Reducing minimum data requirements also reduces a lot of the incentive to actually obtain the data.

If, in the short-term, ACRs continue to be used to derive chronic criteria, careful consideration of the life stages and endpoints used to derive ACRs is needed. As an example, if the chronic toxicity of a chemical is driven largely by mortality of a highly sensitive life stage, then the ACR will likely be highly dependent upon the life stage used in the corresponding acute test. ACRs based on complete testing of the organism's life cycle (e.g., full life-cycle test versus early life-stage test) should be given preference. Differences in acute and chronic MOAs across species need to be carefully considered when selecting ACRs.

Recommendations

- Uncertainty in WQC derivation should be incorporated into WQC expression; regression modeling of chronic data offers a mechanism to quantify uncertainty.

- All WQC derivation methods that extrapolate from relatively small chronic data sets have high uncertainty; emphasis should be placed on developing chronic test protocols and collecting data needed to reduce these uncertainties.

- Different algorithms for deriving chronic WQC have different strengths and weaknesses, but in the absence of comprehensive chronic toxicity data, all derivation procedures are subject to substantial uncertainties. Until the availability of chronic data improves, careful judgment must be used to insure that the weaknesses inherent in each approach do not bias resulting WQC. Values calculated by different methods should be assessed to identify these biases.

Ecological significance of chronic endpoints

Although satisfying from an intellectual standpoint, development of point estimates from chronic exposure–response relations does not, by itself, obviate the problem of deciding what an ecologically significant level of effect should be. Simply put, population size is controlled by 4 processes: 1) births into the population, 2) deaths within the population, 3) immigration into the population, and 4) emigration from the population. Survival and reproduction measurement endpoints from chronic tests address 2 of these processes directly and, as such, provide information relevant to establishing ecologically significant criteria. The growth endpoint may address one, both, or neither of the first 2 processes, depending entirely upon the demographic characteristics of the species and its relationships with other members of the aquatic community. To clarify, species for which gamete production is a function of female size may suffer population-level consequences from reduced growth rate, whereas the reproductive output of species for which such relationships do not hold may suffer no reproductive impairment. Individuals of certain species must attain a minimum size to reproduce at all. Body size also can influence survival during winter starvation or other periods of stress. Consideration of species interactions within the community are relevant when individuals in the population are, for example, prey to size-selective predators. In this case, the significance of growth impairment of the prey species depends upon whether it escapes predation by being smaller or suffers greater predation by not growing to a certain size. The growth endpoint also may have value when the assessment endpoint reflects community dynamics or resource utilization by humans: reduced prey biomass may impact predator populations adversely, and lost fish biomass may be undesirable when that species has economic value for sports or commercial fisheries.

Although typical measurements from chronic tests quantify key demographic parameters of populations, interpreting the ecological significance of these responses to chronic exposure is not straightforward. For example, populations with high reproductive rates ("r strategists") may be able to withstand substantially more reproductive impairment as a result of exposure to chemicals than can those populations producing small numbers of offspring ("K strategists"). On the other hand, species with short life spans may be less able to withstand even temporary reductions in reproduction because the breeding population dies before offspring can be recruited into the population. Populations that experience high survival of early life stages may be more at risk to increased mortality at that life stage than populations displaying low survival at early stages. Thus, the life history characteristics of the species, in combination with site-specific factors, are key determinants of the population-level significance of any effect on demographic parameters.

An approach to resolving the question of significance from a population or community perspective involves the use of extrapolation methods and models directly in development of generic WQC. Along with the understanding of the demographic characteristics of test species, simple models translating endpoints of survival, reproduction, and, as appropriate, individual growth into estimates of population-level responses can be used to predict effects on the demographics of populations of the test species. This approach was used recently by Thursby et al. (2000) to develop draft criteria for dissolved oxygen in saltwater and by Sibley et al. (1997) to assess the influence of nutritional stress on population-level effects in midges.

Alternatively, generic models can be developed that capture the range of life-history strategies employed by species in aquatic systems. These model systems would be used as surrogates for species expected to be present within aquatic systems. The predicted responses of these surrogate "species" might be combined through simple summation, or for added realism, connected in logical networks reflecting predator–prey and competitive relationships, to describe potential effects on communities of species. Delos (1994) recently illustrated some of these concepts, and population models describing several aquatic species have been developed to support this type of approach (Munns et al. 1995). These approaches enhance the relationships between the original measurement endpoints (survival, reproduction, and growth of individuals in toxicity tests) and assessment at the population level and community level.

There are other complications to be overcome. First, many current criterion approaches establish chronic values as the lowest concentration observed to affect any of the 3 measurement endpoints (survival, growth, reproduction). However, effects on multiple demographic parameters often act in concert to influence population growth rate; that is, it is the combination of all demographic effects that determines population-level response. Second, most approaches to data generation for WQC development rely on the responses of specific test species to be representative of the responses expected from other, untested species. While this may be true from a toxicological standpoint, different species with similar toxicological sensitivity may have greatly different life histories, making the population-level consequences of that sensitivity very different across species. In addition, survival, growth, and reproduction within populations may be affected by mechanisms other than direct toxicity as measured in chronic toxicity tests; for example, increased susceptibility to pathogens, parasites, or predation as a result of toxicant exposure can induce population-level effects. Finally, and perhaps most importantly, an improved understanding of density-dependent compensation and recovery processes for populations is needed.

Thus, it is clear that no one level of impairment in laboratory tests can be selected that will represent the same degree of population risk for all species. At the same time, the state of the science is such that organism-specific values cannot be derived for all species of interest in WQC generation. For the present, it is recommend that the Type 1 criteria approach (see Chapter 1) employ default thresholds derived from evaluation of existing information, relying upon best professional judgment to incorporate toxicological and ecological principles. Site-specific refinements of standards (Type 2 criteria) might evaluate the effects of various levels of impairment on the species used to generate the toxicity endpoints, perhaps allowing differences in threshold concentration to be derived as functions of the species' biology and ecology. Even more detailed assessments (Type 3 criteria) could include empirical and simulation evaluations that specifically incorporate the ecological processes operating in the system of interest.

Recommendations

- Chronic tests should continue to employ survival, reproduction, and growth as ecologically relevant measurement endpoints.

- Chronic effect thresholds should be defined using point estimates and other techniques that utilize the full set of data available from toxicity tests.

- Requirements for chronic toxicity data to develop WQC should specify that experimental designs follow a regression approach with sufficient exposure levels to accurately define the concentration causing a specified level of effect.

- For the present, Type 1 criteria should employ default thresholds (e.g., percent impairment) derived from evaluation of existing information, relying upon best professional judgment to incorporate toxicological and ecological principles.

Research needs

- Practical statistical models that define both the effect concentration and the error associated with the estimate.

- Methods to extrapolate toxicological measurement endpoints to population, community, and ecosystem responses.

- Enhanced understanding of compensatory responses and recovery as they influence the long-term effects of chemical exposure.

Time Dependence of Toxicity

As discussed in Chapter 2 ("Exposure Variability: Spatial and Temporal Issues," p 24), biological response to chemicals is determined by a combination of exposure concentration and duration. The interaction of concentration and time at lethal and sublethal exposures can be quite complex but ecologically quite relevant (e.g., Mount et al. 1990; Fairchild et al. 1992; Gagen et al. 1993; Holdway et al. 1994; Barry, O'Halloran et al. 1995). Laboratory and field studies have shown that, on the basis of overall mean concentrations, fluctuating exposures can be more toxic than constant concentration exposures (e.g., esfenvalerate effects on daphnids; Fairchild et al. 1992). Aquatic toxicity tests usually apply constant exposure concentrations over fixed exposure durations, as with 96-hour lethality tests for fish. Thus, comparing laboratory test endpoint concentrations to field concentrations averaged using the length of the test can be inappropriate because greater effects may be elicited by fluctuations occurring in the field.

Criteria are developed from toxicity tests based on the fixed concentration approach but are applied to natural ecosystems where exposures vary in space and time and may have no defined beginning or end. Criteria may be presented in terms of concentration only, or as one or more periods over which monitoring data must be averaged to achieve a limit to both acute and chronic toxicity. For example, U.S. WQC limit the degree of such excursions by applying acute criteria (derived from 48-hour to 96-hour LC50s) to 1-hour averaging periods and chronic criteria (derived from tests 7 days or longer) to 4-day averaging periods. This "one-size fits all" carries the risk of underprotection or overprotection, depending on the compound being regulated. Where no excursions are allowed above criteria concentrations (the one concentration approach), there is a risk of overprotection because the overall allowable exposures are lower than would be the case if the relationship of toxicity to exposure time was recognized in the criteria.

A variety of models have been proposed or tested to account for the interaction of time and concentration, but these often are based only on acute lethality. Where chronic toxicity is derived from a sublethal response, any time adjustment will vary with the species tested, the most sensitive response, and the properties of the chemical that determine pharmacokinetics and toxicokinetics. Models that have been tested include simple or time-weighted averages of concentrations, integration under time-concentration curves, and pharmacokinetic models that predict tissue concentrations under fluctuating exposure regimes in relation to CBR (Hodson et al. 1983; Heming et al. 1989; Hickie 1990; Arthur 1991; Gobas and Zhang 1992; Handy 1992a, 1992b). As reviewed below, none of these approaches are sufficiently well developed to provide a scientifically rigorous basis for incorporating time into criteria development.

Acute toxicity

Chemical concentrations lethal to aquatic organisms decline with time, often showing an exponential approach to an asymptotic value, variously called the threshold, asymptotic, or incipient lethal concentration (Figure 3-4). The time scale of this decline can vary greatly among organisms and chemicals, and the shape can be perturbed by multiple MOAs, delayed mortality, and other factors, but this general shape is still usually observed. Various authors (Mancini 1983; Neely 1984; Chew and Hamilton 1985) have noted that such an exponential decline is expected if the chemical accumulates in or on fish by first order kinetics and if death occurs once a lethal accumulation threshold is reached. The threshold

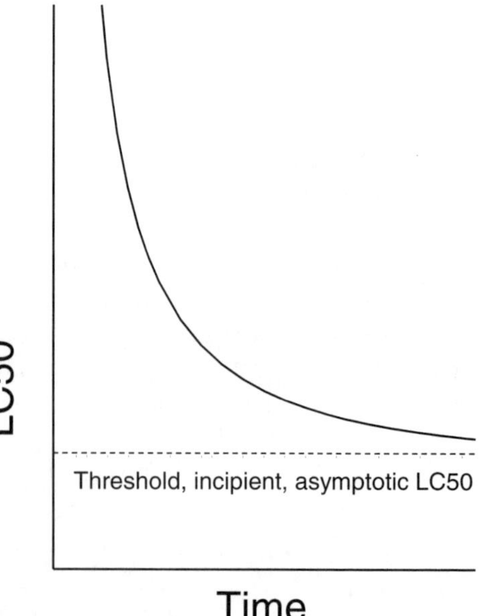

Figure 3-4 Relationship between acute toxicity value and time

lethal water concentration is that which is needed to reach the threshold lethal tissue accumulation (also known as CBR) at steady state, and the rate of exponential decline reflects the kinetic constants for accumulation. Other authors (Connolly 1985; Erickson et al. 1999) have extended this to the concept of accumulation of toxicant damage to the organism, rather than just accumulation of the toxicant, and elimination of damage by repair mechanisms, which produces the same type of equation. If statistical distributions are specified for the variation of model parameters among individual organisms, the models also can predict varying levels of mortality, as well as the effect of time on a specific endpoint such as the LC50.

Mancini (1983) noted that these models not only describe the relationship of lethal concentrations to time under constant exposure but also can be used to predict the lethality of any arbitrary concentration time series. Mancini (1983) and other investigators (Breck 1988; Erickson et al. 1989; Hickie 1990; Meyer et al. 1995) have used this approach to predict the effects of fluctuating concentrations, with mixed results. While predictions can deviate from observed effects, the models still predict the direction and general magnitude of differences between constant and fluctuating exposures. With suitable recognition of uncertainty, these models can provide a framework potentially useful for addressing time issues in criteria.

One such use is setting acute averaging periods on a chemical-specific basis. The 1–hour averaging period in U.S. acute criteria was selected based on observations that, for some chemicals, LC50s for durations of just a few hours did not deviate significantly from those at 96 hours (i.e., the threshold lethal concentration was reached very quickly, and it was imprudent to allow concentrations significantly above the criteria for even an hour). However, this averaging period would be overly restrictive for many chemicals and could substantially restrict the overall exposure concentrations considered acceptable. The toxicity model described above can be used to determine how long the averaging period could be without resulting in more effects than desired, if the nature of fluctuations expected to occur within the period are specified. However, even chemical-specific averaging periods are limited in that they do not completely represent the magnitude and frequency of expected toxicity. One other use for the toxicity model is to eliminate the need for averaging periods completely. Given an expected exposure regime, the toxicity model can predict the expected frequency of various levels of toxicity, providing the basis for a risk-based criterion (Figure 3-5).

The use of such models does have certain limitations and data requirements. Kinetic parameters must be derived for all species of importance to the level of protection of interest to the criteria. Toxicity tests for these chemicals must include monitoring of mortality at a range of exposure durations, and the importance of delayed mortality also must be evaluated. The expected temporal variation of receptor organism exposure must be determined or assumed.

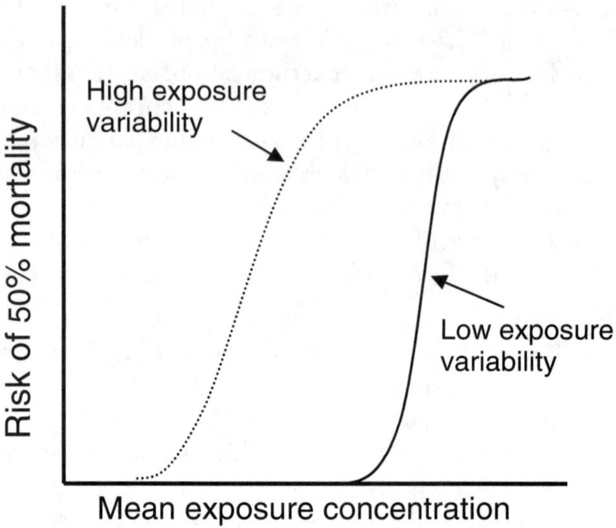

Figure 3-5 Risk curve from application of kinetic toxicity model to exposures with different degrees of variability

Chronic toxicity

The time-dependence of chronic toxicity is not so simply addressed. Chronic mortality is not necessarily an extension of the acute mortality curve. Endpoints such as growth are integrals of growth rates, which are graded responses to chemical accumulation, rather than a simple quantal threshold. Endpoints might depend on exposures to sensitive development stages at only certain times of the chronic test.

The 4-day averaging period for U.S. chronic WQC is based on the need to limit excursions over effects concentrations from the shortest chronic tests (7 day). Even without explicit models for the time-dependence of chronic toxicity, the appropriateness of this averaging period for specific chemicals can be evaluated based on consideration of the nature and duration of the actual chronic tests available for the chemicals and on expected exposure variability. For slowly accumulating chemicals, the kinetics of accumulation of the chemical also can be used to determine minimum durations for averaging periods.

Models for the time-dependence of chronic toxicity can be developed in various ways. Chronic mortality can be evaluated using the same techniques as for acute lethality and possibly even integrated into a single model for mortality. Modeling is easier if accumulation kinetics are known or are much quicker than the toxic response, so that accumulation can be assumed to always be proportional to water concentration. Death, growth, and reproductive rates thus can be evaluated as functions of chemical accumulation. Using these functions and the accumulation model, time series of exposure concentrations can be converted to expected toxicity (e.g., Erickson et al. 1989). The complexities of toxicological responses, such as delays in physiological responses to changes in accumulation, can result in uncertainties, but such efforts might still provide a useful basis for setting chronic averaging periods.

Recommendation

- Emphasis should be given to incorporating chemical-specific information on the kinetics of uptake and toxicity into WQC averaging periods. Acute and chronic toxicity tests designed to provide data for criteria development should be augmented with data describing the time course of each response measured.

Research needs

- Development of reliable models for incorporating time-response data into WQC and risk assessments. Specific needs include
    - determining chemical-specific, acute averaging periods based on kinetic analysis of acute toxicity data;
    - determining chemical-specific, chronic averaging periods based on evaluation of the nature of chronic tests and responses; and
    - conducting studies to improve models for the time-dependence of mortality, to develop models for the dependence of chronic toxicity on exposure duration and variability, and to evaluate the feasibility of risk descriptions that do not use averaging periods.

Bioavailability

The bioavailability of a chemical describes its potential for binding or uptake by an organism. A chemical that is not bioavailable cannot exert toxicity. There now is widespread acceptance of the principle that for many toxicants, only a fraction of the total chemical in water (or sediment or food) is bioavailable and, therefore, capable of causing toxicity. In general, the greater the concentration of bioavailable chemical, the greater the toxicity. For waterborne chemicals, factors that modify bioavailability are generally geochemical and physical properties of the water itself—pH, hardness, alkalinity, DOC, sodium, chloride, and other cations or anions, oxygen status, and temperature. These factors work in different ways: by altering the speciation of the toxicant through complexation or dissociation, by competing with the toxicant for binding or uptake by the organism, by altering the effective permeability of the organism, or by otherwise transforming the chemical to a more or less bioavailable form.

In general, current WQC do not recognize the concept of bioavailability and implicitly assume that 100% of the chemical present in the water column is bioavailable (exceptions are the inclusion of a hardness modifier for some metals and the correction for pH effects on ammonia toxicity). The original data used in developing criteria were usually derived from tests in dilute or synthetic laboratory waters containing minimal levels of factors known to reduce bioavailability. However, these criteria are applied in natural waters where the concentration of bioavailability-reducing factors may be much greater. As a result, the margin of safety is increased and may result in overprotection.

The goal in the future should be to incorporate as many factors that modify bioavailability as is practical into generic WQC (Type 1); by default, they will therefore be incorporated into site-specific (Type 2 and 3) criteria. For many chemicals, there is now sufficient scientific information to incorporate some easily measured bioavailability modifiers into acute criteria by means of equations, models, or nomograms. One major research objective should be to determine how to apply this approach to more chemicals and more modifiers. A second research objective should be to determine how bioavailability defined from acute toxicity studies relates to chronic toxicity.

Cationic metals

Metals for which WQC are commonly developed include copper, cadmium, zinc, silver, lead, cobalt, nickel, iron, mercury, manganese, molybdenum, aluminum, and chromium. At very high concentrations (typical only of spills or other extreme situations), all of these metals kill aquatic organisms by a generalized disruption of the respiratory surface, resulting in suffocation. However, physiological and toxicological research over the past 15 years has clarified the MOA of many metals at concentrations more typical of the environment—that is, in the low µg/L range, at or below the 96-hour LC50s for crustaceans and fish. These are the mechanisms operative at the level of acute criteria. In addition, research has elucidated those aspects of water geochemistry that exert major influences on the bioavailability (and therefore the toxicity) of these metals. Our understanding of the mechanisms of chronic toxicity, and the influence of water chemistry on chronic toxicity, remains incomplete.

Current understanding

Dissolved metals are considered to be the forms of toxicological concern, as particulate metals generally have low bioavailability to aquatic organisms. This perspective was recognized in the U.S. by the 1993 modification to the application of the USEPA WQC for metals (Prothro 1993). The ultimate fate of particulate metals is sedimentary deposition, although the degree to which they may become remobilized into the water column or become available for trophic transfer is uncertain.

Several metals are essential micronutrients (e.g., copper, zinc, cobalt, molybdenum, nickel, iron), and aquatic organisms take them up from the water column through body surfaces. This route is particularly important for algae and unicellular organisms, but uptake from water probably also occurs to a significant extent in fish and crustaceans, especially when dietary sources are deficient. This factor should be kept in mind when deriving chronic criteria to avoid creating deficiencies when protecting against toxicity.

With the exception of aluminum and iron, none of the metals cause respiratory toxicity in fish and invertebrates except under extreme conditions (e.g., spills). The acute toxicity of most metals appears to be related to the binding of the metal cations to anionic sites on or in the gills. These anionic sites are likely proteins that play critical roles in the ionoregulatory transport functions of the gills. For example, copper, silver, mercury, and aluminum (at lower pH) specifically inhibit Na^+ and Cl^- uptake processes, whereas cadmium, zinc, cobalt, and manganese specifically inhibit Ca^{2+} uptake processes (Wood 1992; Markich and Jeffree 1994; Heath 1995; Bergman and Dorward-King 1997; Playle 1998). At slightly higher concentrations, metal cations exert a common effect to increase the permeability of the gills to electrolytes, thereby accelerating diffusive losses of Na^+, Cl^-, and other major plasma ions. The effect appears to involve displacement of Ca^{2+} from the proteins that stabilize the junctions between gill cells. As a result, the paracellular pathways become increasingly "leaky." For most metals, the internal physiological consequences of these deleterious surface effects (e.g., osmoregulatory failure, hypocalcemia) are far more serious than any internal effects resulting from the entry and accumulation of metal.

In fresh waters, the most important aspects of water geochemistry affecting the bioavailability and toxicity of metals are pH, DOC, alkalinity, Na^+, Cl^-, and hardness (or more explicitly, calcium concentration because magnesium concentration, the other component of water hardness, is much less potent). In brackish and marine waters, salinity becomes an important influence. Changing pH alters the proportion of cations that are free as well as altering H^+ competition with cations for binding sites. The influence of DOC is via complexation of the metal, while alkalinity (bicarbonate) forms hydroxide and carbonate complexes; complexes also are formed by Cl^-. Na^+ and hardness (Ca^{2+}) effects involve competition for binding sites. Increasing salinity may involve all of these effects, though the much higher Ca^{2+}, Na^+, and Cl^- concentrations in seawater are dominant influences. For this reason, the acute toxicities of most metals tend to be lower in seawater than in fresh water.

BLM

Recently, a BLM has been developed to represent the interaction of metals with respiratory surfaces and the ensuing toxicity. This model evolved from the original theoretical frameworks for competition and complexation developed by Pagenkopf

(1983) and Morel (1983), and the measurements of metal binding to gills by Playle et al. (1993a, 1993b). The model approach is presented in detail in Bergman and Dorward-King (1997) and by Playle (1998). In brief, the model focuses on the initial step in metal uptake as the binding of the free metal cation to ligands on the respiratory (gill) surface. These sites have discrete affinities that can be described by conditional equilibrium stability constants (K_D) and maximum site numbers or binding capacity (B_{max}) for the metal cations; these values have been characterized recently for some metals (copper, cadmium, silver, cobalt, zinc, nickel) in some organisms (rainbow trout, fathead minnows, *Daphnia magna*). By linking this binding to generic geochemical modeling programs, the model predicts the degree to which metal-binding sites on the gills are saturated with metal. The advantage of this approach is that it predicts the amount of the metal ion that is bound to the target surface rather than the free metal ion concentration in the water column. Death is assumed to occur when the amount of bound metal reaches a critical value.

By incorporating K_D values for H^+, Ca^{2+}, and Na^+ binding to these same sites into the model, the effects of competition are explicitly considered. By incorporating K_D and B_{max} values for the metal binding to DOC, Cl^-, carbonate, and other anionic ligands of interest, the effects of complexation are explicitly considered. Finally, by incorporating comparable values for other metals, this approach has the potential to predict the effects of metal mixtures. Critical here is knowledge as to whether different metals bind to the same or different anionic sites on the gills, for which physiological tests can provide important indications. Information available to date indicates that those metals that interfere with Na^+ and Cl^- uptake bind to a different population of sites than those that inhibit Ca^{2+} uptake.

At present, one limitation of the BLM is its failure to address kinetics of accumulation and toxicity. Longer-term metal binding kinetics are poorly understood. For example, during chronic exposures, organisms may accumulate gill metal burdens well above saturation of all possible high affinity sites, yet toxicity does not occur (e.g., Hollis et al. 1999). A second limitation is that the database relating short-term binding to acute toxicity is still fairly small and that for chronic toxicity smaller still. Nevertheless, the limited data available indicate a direct, strong correlation between the calculated percent saturation of gill binding sites (at 2 to 24 hours) and acute toxicity (96-hour LC50s) within single-species tests.

Incorporation into WQC

Current USEPA WQC for metals incorporate only one of the important factors now known to modify bioavailability—that is, hardness—and do so by means of empirical equations. Because these equations fail to separate the effects of calcium from those of magnesium, the relationships are uncertain. In addition, these relationships incorporate not just the effects of the hardness cations, but any ligands correlated with hardness, increasing uncertainties if these correlations are not similar in the waters to which the criteria are applied. For example, some hardness relationships reported for silver have recently been demonstrated to likely be due to chloride (Wood et al. 1999). For some metals (copper, cadmium, silver, nickel), there is now sufficient toxicological information available to separately incorporate important influences on bioavailability—pH, DOC, alkalinity, calcium, and selected other factors—into Type 1 criteria by means of multifactorial equations. Some such equations already exist (e.g., Erickson et al. 1987; Meador 1991; Welsh et al. 1993). The availability of generic geochemical modeling programs has greatly simplified the computations needed. Furthermore, the opportunity exists through the BLM to incorporate the receptor organism itself into this framework.

The BLM is currently focused on water column exposure. Though its applicability to sediments has been proposed, it has not been rigorously tested. Potential effects mediated through food-chain transfer of metals also are not considered (see "Dietary Concern in the Development of WQC," p 96).

Recommendations

- For all metals of concern, Type 1 acute criteria should attempt to incorporate pH, DOC, alkalinity, calcium, and other appropriate factors by means of multifactorial equations or computer models.
- The geochemical-biological models, such as the BLM, provide a reasonable basis for directly incorporating bioavailability into criteria derivation for copper, cadmium, silver, nickel, and possibly other metals.
- Chronic toxicity data should be generated to determine the applicability of bioavailability modeling approaches to chronic toxicity.

Research needs

- Information on how to translate the output of the BLM and other models, typically based on data from fathead minnows, rainbow trout, and daphnids, to other potentially sensitive species; to better understand the time-dependent component of the gill binding versus toxicity relationship and the relationship of gill binding to chronic toxicity.
- A more complete understanding of the applicability of the BLM to a wider variety of organisms, particularly invertebrates that use different respiratory strategies and structures.
- An exploration of the links between exposure in the water column, sediment, and diet and the applicability of those links to WQC.

Organometallics

Some metals can exist as organic compounds that occur in, or are even formed in, the environment (e.g., methyl mercury, tributyl tin, ethyl lead). These compounds differ from the metals discussed in the previous section because they exist as neutral, less polar compounds in the aquatic environment. As such, they do not interact with anionic ligands on or in the gill surface, but rather diffuse directly through the respiratory epithelium or body surface into the organism, much in the same manner as nonionic organic molecules. Toxic mechanisms are generally internal; rarely are waterborne concentrations sufficient to cause nonspecific gill damage. While some metabolism and excretion may occur, trophic transfer and bioaccumulation are a major concern for organometallics. DOC will bind organometallic compounds, and hardness has been shown to reduce uptake, probably via the general stabilizing action of calcium on the permeability of surface. Effects of pH appear inconsistent, and there are limited data on the effects of water geochemistry on uptake rates of these compounds. Their toxic effects occur most often at higher trophic levels. Hence, there does not appear to be compelling evidence at this time to incorporate bioavailability-modifying factors into WQC for organometallic compounds. Rather, they should be treated as unique families of compounds, unrelated to inorganic metals, and each with its own exposure and toxicological characteristics. Because many of these compounds sorb to sediments, corresponding WQC must simultaneously consider sediment as a source of environmental exposure.

Recommendations

- Incorporation of aqueous bioavailability considerations into WQC for organometallics may not be warranted.
- WQC approaches for these compounds must simultaneously consider sediments.
- Residue-based criteria may be effective for organometallic compounds.

Metalloids

The metalloids, arsenic, selenium, and boron, share the characteristics of having more than one inorganic oxidation state and existing in anionic forms and/or organic complexes. For example, selenium is commonly found in the environment as selenate (SeO_4), selenite (SeO_3), and selenomethionine, in order of increasing toxicity to aquatic organisms. This polymorphism greatly complicates assessing their environmental behavior and toxicity and has introduced substantial controversy into attempts to assess their environmental risks.

Most metalloids are common constituents of minerals and soils and may be present at potentially toxic concentrations even in relatively undisturbed environments. Activities that increase soil leaching, such as irrigation or mining, can cause significant elevations in concentrations of these chemicals in the aquatic environment. Anthropogenic sources also may be of concern. Arsenic and boron are both used for wood preservation and as pesticides. Arsenic, boron, and selenium can be released through mineral extraction, mineral processing, or coal combustion. One concern for selenium is that it is an essential nutrient, and the range of exposures between deficiency and toxicity appears to be very narrow.

Speciation of metalloids in the environment is controlled largely by redox conditions, pH, and in some cases bacterial action (e.g., methylation). Because metalloids can exist in both organic and inorganic forms, their environmental behavior and potential for accumulation by organisms can vary widely. Toxicity also appears to be highly dependent on speciation. The similarity of some metalloids to important physiological elements (e.g., sulfur, phosphorus) yields some very specific toxic mechanisms. Boron is particularly toxic to plants (Gupta et al. 1985). Accumulation and concentration in the food chain may be of concern for some metalloids, with selenium as the most notable example; effects of selenium accumulation may be most pronounced in wildlife consumers of aquatic biota and plants, although some fish species that consume benthos seem to be sensitive as well.

Because of the complex speciation and behavior of metalloids, WQC for these chemicals must be carefully designed to recognize the specific environmental pathways and toxic mechanisms applicable to each chemical, including consideration of accumulation and subsequent effects in wildlife and humans. Integration of sediment criteria with WQC may be necessary to characterize risk adequately. Because of concerns for accumulation, residue-based criteria may be a viable approach for some metalloids, but only if they can effectively handle the varying toxicities of different chemical forms.

Recommendation

- To improve criteria derivation and monitoring procedures, a better understanding of the speciation of metalloids is required, at least to the level of inorganic versus organic complexes. This applies to both water-based and sediment-based criteria as well as criteria based on tissue residue.

Nonionic organic compounds

For nonionic organic chemicals, the primary water chemistry variable affecting bioavailability appears to be organic carbon. In water, hydrophobic chemicals associate with both particulate organic carbon (POC) and DOC. While bound chemicals are measured by many analytical methods, binding to both POC and DOC reduces the toxicity of those chemicals in water column exposures (Gobas and Zhang 1994). The influence of POC can be isolated by measuring only the dissolved fraction of the chemicals (through filtration or centrifugation) or by calculating the dissolved concentration based on the chemical's organic carbon partition coefficient (K_{oc}). Chemicals bound to DOC are still measured as dissolved, although experimental data suggest that it is only the freely dissolved portion (i.e., not bound to DOC) that is toxicologically available in the water column (Gobas and Zhang 1994). Partition coefficients for DOC (K_{DOC}) can be used to estimate freely dissolved concentration based on the measured concentration of dissolved chemical, DOC, and K_{DOC}. K_{DOC} is generally thought to be roughly one-tenth of K_{ow} (USEPA 1995b).

Under current USEPA WQC procedures, DOC binding is not considered, and binding to POC is accounted for only if chemical measurements are made on a dissolved basis. With regard to effects assessment, toxicological data are selected to exclude toxicity data derived in waters having elevated DOC or POC. In this way, the criteria database is biased toward toxicity reflective of freely dissolved concentrations, although some degree of POC and DOC binding is possible even in these relatively "clean" waters.

In practice, many natural waters contain concentrations of POC and DOC sufficient to affect bioavailability. As indicated, this effect is ignored in the generic application of WQC, at least with respect to the dissolved fraction. In theory, this effect could be accounted for in a Type 2 criterion through the use of a WER (USEPA 1994), though studies describing WER application to nonionic organic compounds have not been published. Procedures for separating DOC-bound and freely dissolved chemical have been published (Landrum et al. 1984; Burgess et al. 1996), but they have not been widely applied or validated in the context of WQC. At DOC concentrations typical of most surface waters, DOC binding will account for a toxicologically meaningful percentage of dissolved chemical only for chemicals with log K_{ow} of 5 to 6 or above (Gobas and Zhang 1994).

Although procedures exist for correcting water concentrations for DOC binding, it is critical to do so only in recognition of other issues pertaining to high K_{ow} organic chemicals. The same properties that enhance POC and DOC binding for these chemicals in the water column also increase their potential to accumulate in sediments, organisms, and their prey items. Particularly where the chemical activity of organic chemicals in sediment is higher than that in the water column, transfer of chemical through the food chain may be a substantial route of exposure not explicitly considered in most current procedures for WQC development. Refinement of bioavailability assessment for high K_{ow} organics should proceed only in concert with an expanded consideration of fate (sediment) and exposure (diet).

Recommendation

- Continue to base WQC on data derived from toxicity studies using water with low POC and DOC. For high K_{ow} compounds, it is recommended that the coherence of WQC increased with sediment and dietary routes of exposure, possibly through a residue-based criterion. It is also necessary to evaluate the potential of using K_{DOC} and/or WER studies to account for DOC binding.

Weak acids and bases

Various toxicants, including ionizable organic chemicals (e.g., chlorinated phenols) and various nonmetal inorganic chemicals (ammonia, sulfide, cyanide) are weak acids or bases that exist in both unionized and ionized forms. The relative amounts of each form can vary substantially over the pH range of concern to WQC. Permeability of biological membranes is generally much greater for the unionized form, resulting in bioavailability (and therefore toxicity) that varies with pH. However, toxicity generally cannot be attributed solely to the more bioavailable form, so criteria cannot be expressed as a constant in terms of this form. For example, although ionized ammonia can cross gill membranes much less readily than unionized ammonia, it is present in much greater concentrations, especially at lower pH, and may be a significant contributor to ammonia toxicity despite its low bioavailability (Erickson 1985). Despite its charge, the ionized form of pentachlorophenol appears to partition somewhat into organic phases (Kaiser and Valdmanis 1982; Jafvert et al. 1990) so that it should cross biological membranes. This is likely the reason why measured toxicity exceeds the toxicity expected from the unionized form (e.g., Saarikoski et al. 1986; Kishino and Kobayashi 1995). A further complication is that reduced pH at the gill surface can alter the relative amount of the ionized and unionized forms, thereby changing uptake in a manner not reflective of the relative amounts of the forms in the "bulk" exposure water. At a minimum, the dependence of the toxicity of weak acids and bases on bulk water pH should be established to develop WQC suitable for such chemicals, but other physicochemical factors also can affect toxicity. For example, cyanide toxicity may be affected by complexation with various metal cations, and ammonia toxicity can be altered by the ionic composition of the test water, especially at low pH (Ankley et al. 1995; Borgmann and Borgmann 1997). Consideration of the aquatic chemistry of the toxicant and its possible relationships to toxicological responses can help identify such factors.

Recommendation

- For ionizable compounds, effects of pH on toxicity should be incorporated into WQC development.

Composition of the Toxicological Database

Although specific procedures vary, all WQC approaches rely primarily on laboratory toxicity data, and use some form of extrapolation or interpolation to estimate a chemical concentration that will protect most species, most of the time. The assumption is that, from a subset of species, a concentration can be calculated that will be reasonably protective of the majority of species both tested and untested. This extrapolation is most often done by fitting a statistical distribution to the available toxicity data, and then projecting the concentration associated with the desired level of protection (e.g., 95% of species). The uncertainties in this approach stem primarily from 2 issues: 1) the influence of the database size on the prediction, and 2) the degree to which the tested species represent the sensitivity of all species that the WQC is intended to protect.

These uncertainties involve more than statistical extrapolation alone. Because the selection of species for toxicity testing is influenced by a number of practical and historical factors, there is no reason to assume that these species represent a typical subsampling of all organisms. In addition, many WQC derivation procedures have requirements for certain taxonomic groups that must be represented in the database. For example, in the USEPA guidelines, 2 of the requirements are for planktonic crustacean (e.g., daphnid) and salmonid fish. Both of these requirements are driven by the observation that one or both of these species are often among the most sensitive of species for a given chemical. The result is a bias in the toxicity database relative to the expected distribution of organisms as a whole.

Evidence for this bias can be found through analysis of existing databases for USEPA criteria. If an FAV (95th percentile of the LC50 data) is calculated using only data for the 8 required taxa, it will almost always be lower than the FAV calculated using all available data. If the minimum database was a random representation of all data, then one would expect that the 2 values would tend toward equality, sometimes higher and sometimes lower. Instead, the lower values computed from the minimum database indicates that this data set is biased toward more sensitive species, relative to the overall data set. Uncertainty related to data bias is much more difficult to quantify than simple statistical uncertainty. From existing WQC, the direction of this bias appears to be toward a higher level of protection than indicated by the confidence level of the statistical analysis, at least as represented in the USEPA guidelines. It is for this reason that the USEPA criteria statements are relative to "95% of the tested species," rather than to 95% of species in general.

Numbers of species tested and the range of sensitivity

Water-quality criteria development often specifies acute and chronic toxicity data from a certain minimum number of test species to estimate a range of sensitivity and to ensure protection of the most sensitive species. The species sensitivity curves also provide a basis for probabilistic risk assessment, so that the consequences or risks of exceeding the criteria can be defined quantitatively. Clearly, the confidence in each criterion and its associated species sensitivity curves will increase with the numbers of species tested (Figure 3-6), but the influence of the numbers of species on the uncertainty of the criterion is neither expressed nor considered when interpreting the risk associated with exceeding the criterion. Further, for compounds for which there is a great deal of concern but few data, no criteria may be available to aid risk assessors and managers (see "Determining Approach and Data Needs for Effects Assessment," p 54).

One solution to these problems is to calculate the uncertainty or variance associated with the estimation of each criterion. For the users, it will be immediately obvious that some criteria are more reliable than others, in the sense that some are supported by more data than others. For relatively new and/or poorly studied chemicals, the variance or uncertainty of the criteria would be relatively large, while the opposite would be true of well-studied chemicals such as copper. The overall result will be more realistic risk assessments, the inclusion of uncertainty into decision-making, and the appreciation of the potential for over- and underprotection. During implementation, these uncertainty limits could be incorporated into risk assessments for site-specific criteria and recognized in the interpretation of monitoring data.

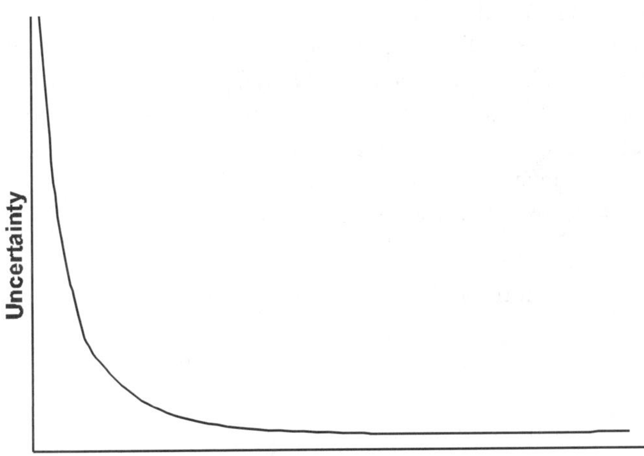

Figure 3-6 Hypothetical relationship between uncertainty in WQC and the number of species tested

Another approach might be to create another class of criteria for chemicals with less than the minimum data requirements. The criteria for such chemicals would be more uncertain and could be described as provisional guidance and could incorporate safety factors that reflect expected inaccuracies. An approach like this was used to create 2 tiers of criteria in the U.S. Great Lakes Initiative. Because WQC often are used for applications other than water-quality standard development, these provisional guidelines would provide access to existing information and a head start to developing site-specific recommendations. Relative to current uses of WQC, the challenge in this approach would be to ensure that provisional guidelines are used only in applications compatible with their high uncertainty.

Recommendations

- The statistical uncertainty associated with WQC and species sensitivity curves should be expressed as part of each criterion.
- The requirements for a minimum number of test species and chronic toxicity values could be reduced to make data and guidance available at an earlier stage, but the resulting uncertainty in criteria supported by limited chronic data must be considered prior to development of standards.

Taxonomic coverage

Existing WQC derivation procedures are driven largely by available toxicity data on a few fish and invertebrate species and suffer from the uncertainty as to whether these groups of organisms are sufficiently representative of the entire aquatic ecosystem. Though many WQC guidelines require certain taxonomic coverage, some taxonomic groups are poorly represented in most toxicity data, often because the species are difficult to obtain or test, or because standard methods do not exist for these species. The result is that the species distributions in toxicological databases are not random, but actually reflect a number of practical, logistical, and historical influences. In practice, many WQC derivation procedures assume, or appear to assume, that the data are random. It should be noted that some suggestions for focusing the data collection process (e.g., focusing data collection on suspected sensitive taxa) will increase bias in the distribution of toxicity data and will require careful consideration during data analysis.

The USEPA WQC guidelines attempt to ensure reasonably broad taxonomic coverage by placing minimum taxonomic requirements on the database (fish, crustacean, insect, etc.). However, even these requirements leave room for important taxa to be absent. As is noted below, coverage of plants is minimal, with the exception of unicellular algae. An important aspect of this problem is the issue of threatened and endangered species, which would be addressed by site-specific (Type 3) criteria; this is discussed further in Chapter 4.

Animal taxa

With regard to animals, recent attention has been focused on improving representation of amphibians and certain mollusks (e.g., unionid mussels). Amphibians are attractive test species because their external metamorphosis and transition to a terrestrial form involves distinctive and major development events that are subject to stressors present in the aquatic environment, including chemicals. Field studies have shown that various species of amphibians are in decline or are already extinct (Corn 2000). Recent observations also have shown high proportions of deformities among certain populations of frogs and toads (Ouellet et al. 1997). Although the causes of both deformities and population declines remain uncertain, toxic chemicals are a possible cause, and current WQC development procedures generally do not include measurements of potential effects on amphibians. An impediment to evaluating amphibians is the lack of standard test methods. Other than *Xenopus laevis* (the African clawed frog), few species have been routinely used in laboratory studies.

It is generally believed that the early life stages of freshwater mussels are the most susceptible to toxic chemicals, and successful techniques have been developed for testing with glochidia of unionid mussels (Johnson et al. 1993). Glochidia are markedly more sensitive to pesticides and organic compounds than are juvenile or adult mussels, but less sensitive than some standard cladoceran and fish test species (Keller 1993; Keller and Ruessler 1997).

In addition to amphibians and mollusks, insects often are poorly represented relative to their diversity in the aquatic environment. For chronic endpoints, there are difficulties in handling many species in the laboratory and in testing organisms with nonaquatic life stages. Even for those species that have been tested, many studies focus on later developmental instars that are thought to be less sensitive than early instars. Nonetheless, life-cycle test methods have been developed for certain midge species (Pascoe et al. 1989; Benoit et al. 1997).

Of greater significance is the large number of marine taxa that are rarely, if ever, tested in the laboratory, even for acute toxicity. Corals are just one of many marine invertebrates that are rarely represented in the toxicological literature.

Recommendations
- Test methods are needed for taxa that are currently underrepresented.
- Minimum data requirements in existing WQC programs should be carefully evaluated to insure adequate species coverage for the intended uses of the criteria.

Effects on plants

Current criteria, for the most part, do not rigorously address phytotoxicity. Algal toxicity studies have been used to a limited extent, but their current use suffers from poor test design and interpretation. Standard methods for toxicity testing with microalgae (e.g., American Society for Testing and Materials [ASTM] 1998) generally involve preparing test concentrations in growth media relatively low in nutrients, inoculating with algae, and measuring growth over a period including the exponential growth phase (e.g., 96 hours). Typical exposures are static, as methods for flow-through algal tests (e.g., using chemostats) have not been widely used, although microplate methods have been developed and have shown increased sensitivity to metals compared with static exposures (Radetski et al. 1995).

Because algal populations increase exponentially during the course of the test, algal toxicity tests should be considered as chronic and not acute tests. An outdated perception that algal species are not sensitive to most toxicants has been superseded by information demonstrating high sensitivity to a variety of chemicals (Walsh et al. 1982; Benenati 1990; Lewis 1995). Though potentially very sensitive, algal test results can be heavily influenced by test design. For instance, tests of metals must consider that many are essential nutrients for algal growth and must be present in small amounts in the test medium, along with chelating agents to prevent precipitation. Excessive amounts of chelating agents, however, may complex the metal of interest and reduce bioavailability. Furthermore, algal strains may be adapted to different nutrient concentrations, resulting in varying sensitivities to the chemical of concern (van Tillborg and van Assche 1998). Some studies also have indicated different responses of algae when tested at the much lower densities typical of natural systems. Natural assemblages of phytoplankton sampled in Great Lakes water showed reduced productivity as indicated by uptake of ^{14}C CO_2 during exposure to mixtures of metals or to suspended contaminated sediments (Wong et al. 1978; Munawar et al. 1989). Algae also excrete compounds that can complex metals (van den Berg et al. 1979), so that complexation capacity of test systems may change with the density of algae during the test. Clearly, media composition, metal speciation, inoculum source and density, and other test parameters must be carefully considered as a part of the test design.

The results of algal toxicity tests are usually expressed as a median effective concentration (EC50) for growth inhibition. However, the ecological significance of 50% inhibition of growth in an exponentially growing algal population (or any other rapidly growing organism) is uncertain. As long as a few algal cells remain viable, the population has the potential to recover rapidly. Therefore, algal assays designed to determine the algistatic and algicidal concentrations (Payne and Hall 1979; Hickey et al. 1991) may be useful in assessing recovery, particularly where exposures are of limited duration and frequency. Under the same chemical exposure, differential growth inhibition among species may cause changes in phytoplankton population dynamics. The ecological significance of these changes is not easily determined. Measurement endpoints that address phytoplankton function (e.g., chlorophyll content, carbon fixation) may aid the interpretation of algal test data.

Because the sensitivity of algal species to phytotoxic chemicals can vary significantly, tests with species representative of different classes are useful. For example, the algal classes mentioned in current guidelines for pesticide registration in the U.S. are green algae, blue-green algae, and diatoms (Holst and Ellwanger 1982). However, because algal test methods may be relatively inexpensive and readily adaptable to new species, caution is necessary to ensure that algal data do not inappropriately dominate chronic data sets within WQC development.

The assumption that microalgae are representative of all aquatic plants has not been verified. Indeed, with the phylogenetic and morphologic diversity among "aquatic plants" (e.g., cyanobacteria, microalgae, macroalgae, submerged vegetation, emergent vegetation), it is unlikely that tests of unicellular green alga would always be protective of all aquatic plants. Toxicity tests with duckweeds (family Lemnaceae) have been used to provide additional phytotoxicity information, which is certainly desirable. Duckweed toxicity test methods are well developed (ASTM 1991) and their use is recommended; however, the assumption that duckweeds are representative of higher aquatic vascular plants remains unproven. There is a strong need to identify appropriate aquatic macrophytes to include in criteria derivation and to develop and validate test methods for these species.

Recommendation
- Algal toxicity data should be included as part of a minimum data set for WQC derivation. Algal toxicity data are easily generated, but caution is needed because algae may not be representative of all aquatic plants. The assessment of recovery potential (algistatic versus algicidal effects) and the assessment of functional endpoints may be valuable.

Research needs
- Test methods are needed that appropriately address ecologically relevant endpoints for macrophytes.

Use of nonnative species

Some WQC development schemes restrict toxicity data to those species native to the jurisdiction. This requirement may be unnecessarily restrictive for the process of deriving generic (Type 1) criteria and may actually increase, rather than decrease, uncertainty. Most WQC are intended to protect "most of the species, most of the time" and, as discussed previously, use the distribution of toxicity data to estimate concentrations that will meet this objective. For this reason, increasing the depth and breadth of species representation in the toxicological database should better estimate the distribution of species sensitivity.

The presumed reason for excluding nonnative organisms is to avoid potential bias from organisms that may occupy unusual habitats, or are physiologically different from native species. There may be little biological justification for this separation. For example, a recent comparison between tropical Australian and temperate North American data for the toxicity of copper to freshwater crustacea and fish concluded that there were no significant differences, after normalizing for major differences in water chemistry (Markich and Camilleri 1997). In this analysis, the differences among taxonomic groups were generally greater than regional differences among species within the same groups. Hence, the risk of introducing nonrepresentative data may be small compared to the benefits of strengthening the overall database. The current U.S. approach already makes use of species that do not occur in many regions of the U.S. Data from nonnative species may provide insights into taxonomic groups that have particularly high sensitivity to a particular chemical; other members of these groups may be present within the jurisdiction, but not included in the toxicological database for a variety of reasons (e.g., too rare or too difficult to test).

Bias from unusually sensitive nonnative species could be removed easily by reevaluating the applicability of data from such species when they exert a strong influence on the criteria derivation (e.g., present in the tails of the distribution). Further, site-specific (Type 2) criteria could be derived by using only species native to a particular region, as described for the existing "recalculation" procedure included in the USEPA guidelines for site-specific WQC (Carlson et al. 1984). For generic criteria, however, restricting the analysis to native species might limit the evaluation needlessly and may further jeopardize the implicit assumption of a representative distribution of species sensitivity.

Recommendation

- Data from nonnative species should be used in derivation of Type 1 criteria.

Combined use of marine and freshwater data

A similar argument can be constructed for combining freshwater and marine data. The benefits of increasing the database size by pooling the data may outweigh the risks of introducing nonrepresentative data. The potential for differences between marine and freshwater organisms could be tested statistically, although the relatively high variability and limited number of toxicity test data within taxonomic groups may make these comparisons weak. Significant differences may be expected for certain toxicant classes, such as metals, for which speciation, bioavailability, and the physiological responses of organisms might differ between fresh and saline waters. However, for many other contaminants, bioavailability and speciation issues may be less influenced by salinity; in these cases, physiological sensitivity may be assumed to be comparable between freshwater and marine organisms, and

WQC could be developed using a larger, combined data set. This suggestion would apply primarily to generic criteria; for site-specific criteria, the database would presumably be narrowed to reflect only site-specific taxa, and the criteria and confidence intervals revised.

Recommendation
- Data from freshwater and marine toxicity tests should be pooled for derivation of Type 1 criteria where chemical speciation and physiological interactions of seawater with toxicity are expected to be small (e.g., nonionic organic chemicals, but not metals).

Data-quality considerations

Effects data used for criteria derivation should undergo careful scrutiny to determine compliance with minimum standards of acceptability. Existing USEPA guidelines contain general guidance on quality issues, but there has been recent effort to develop a more specific, standard evaluation procedure for data quality (personal communication, CE Stephan, USEPA, Duluth, Minnesota, USA). Critical evaluation points are reasonably well established for the more common types of toxicity tests and test species, but less so for others. In some situations, data from a single study may drive criteria to low concentrations, particularly for data in the tails of the distribution (either high or low sensitivity); this is aggravated when data availability is limited. The more influential a particular test result is in the derivation of a criterion, the more important it is that the test methods and data analysis conform to accepted practice. Further, because of the uncertainty inherent to an individual study, replication of a critical study, or additional studies to augment the database, should be pursued.

Recommendations
- Standardized data evaluation guidance should be developed.
- In cases where one study heavily influences the criterion value, that study should be replicated or otherwise corroborated.

Mixture Implications for WQC

Most existing WQC consider chemicals singly, rather than as mixtures; for classes of chemicals that routinely co-occur as mixtures, such as cationic metals, polycyclic aromatic hydrocarbons (PAHs), and PCBs, this approach clearly has the potential for misrepresenting toxicity.

Extreme antagonism or synergism is rarely observed in environmental mixtures, so most mixture approaches assume additive, or close to additive, interactions. Two conceptual models have been developed to described additive toxicity: response addition and concentration addition (Broderius 1991). In the response addition model, biological responses to chemical exposure are assumed to be additive, even if the responses are elicited through different MOAs. The concentration addition model describes situations where chemicals that operate through a similar MOA combine additively (normalized to potency) to exert adverse effects. The conceptual underpinnings of the response addition model are difficult to test, and as a consequence, this approach has not been widely accepted as a technically viable method in quantitative risk assessment. In contrast, the concentration addition model has been evaluated successfully in aquatic species for a variety of MOAs including narcosis (Hermens et al. 1984; Broderius and Kahl 1985; Swartz et al. 1995, 1997), photoinduced toxicity (Erickson et al. 1999), and various receptor-mediated pathways (Norberg-King et al. 1991; Zabel et al. 1995; Bailey et al. 1996; Walker et al. 1996). Based upon these observations, we believe that technically-defensible mixture criteria can be derived using a concentration addition model for groups of chemicals with a similar MOA.

This approach, embodied in the toxic equivalent concentration (TEC) concept, has proven quite effective for PCBs, as well as for PCDDs and PCDFs, all recognized as AhR agonists. (Cook et al. 1993, 1997). The basic method derives potency factors for individual PCDD-PCDF-PCB congeners for biological endpoints known to be associated with stimulation of the AhR; these values, termed "toxic equivalency factors" (TEFs), are expressed in terms of their potency relative to TCDD. To derive a value indicative of mixture toxicity for a particular sample (generally tissue), congener-specific TEF values are multiplied by their concentrations in the sample, and these values are summed to provide a total TEC. This value is then compared to toxicity thresholds for TCDD to predict risk.

There are several data requirements, assumptions, and associated uncertainties implicit in the generation of criteria using the TEC approach. First, the effects of the chemical must be defined relative to a specific biological effect; in the case of TCDD, this is generally mortality of early life stages of fish (Cook et al. 1997). Second is the assumption that a concentration addition model is appropriate for predicting chronic toxicity; this appears reasonable with respect to early-life stage mortality in fish (Zabel et al. 1995; Walker et al. 1996). Third, it must be assumed that the relative potency of individual congeners is comparable between the species of interest and the test system used to generate TEF values. Where interspecific variability is high, use of generic TEFs across species may be inappropriate, particularly for the derivation of universally applicable national criteria. Finally, because the TEC approach is often based on tissue residues (see CBR discussion in "Residue-based criteria," p 55), it is necessary to have appropriate "translators" (models) to convert tissue concentrations to water and/or sediment concentrations for the purpose of regulation or remediation. This is a challenge common to any tissue residue-based criterion, but is particularly problematic for PCDDs, PCDFs, and PCBs, which rarely can be measured with any degree of reliability in water samples.

The TEC approach also has potential for understanding the joint toxicity of PAHs (see Swartz et al. 1995). For the large number of nonpriority PAHs (including oxy-PAHs), quantitative structure-activity relationship (QSAR) analysis may provide a basis for estimating relative potency for inclusion in TEC analysis (Swartz et al. 1995).

The BLM currently being developed to predict bioavailability of metals (Bergman and Dorward-King 1997; Playle 1998; Wood et al. 1999) may also apply to metal-mixture interactions. Once gill-metal stability constants for a large number of metals have been determined, it should be possible to determine the relative affinity of metals for gill binding sites. This will provide a basis for predicting the relative contribution of different metals in mixtures to toxicity under varying conditions of water quality.

Recommendation

- For chemicals with similar MOAs, WQC development should consider the interactive toxicity of mixtures.

Research needs

- The concentration addition model should be tested for several MOAs, particularly for chronic endpoints.
- TEFs should be developed for different chemical classes and toxic mechanisms of concern.

Criteria for the Protection of Wildlife

Nonaquatic wildlife species may be exposed to waterborne chemicals either through ingestion of water or sediment, or by ingestion of prey items containing chemicals originating from the water. Gulls, cormorants, shorebirds, waterfowl, and some passerines are examples of birds that use various forms of aquatic life as a source of food, while mammals such as mink, otter, and sea lions, and reptiles such as alligators and turtles also are exposed to contaminants from aquatic sources. For example, consumption of fish containing high concentrations of PCBs or dioxins has caused reproductive failure, morphological abnormalities, or even mortality in mink, river otter, and bald eagles (Bradbury 1996; O'Hara and Rice 1996). Therefore, in addition to WQC that protect aquatic life, WQC are required to insure that wildlife do not receive toxic exposures to chemicals through water, sediment, or prey.

If the aquatic criterion is insufficient to prevent loss of organisms consumed by wildlife, indirect effects also may result from a reduction in food quantity or quality (USEPA 1998). An example would be the loss of snails and other mollusks that are used extensively as a food resource by scaup (Bellrose 1976). Crayfish in streams are preyed upon heavily by mink at certain times of the year, and their absence could affect reproduction of mink. Survival of these food resources must be protected through the aquatic life criteria if the goal is to manage for sustainable wildlife populations.

Residue criteria

Wildlife criteria should be based on the concentration of contaminants in aquatic prey or on critical tissue levels within the wildlife species of concern. Because these involve different approaches, they will be discussed separately. Nevertheless, either of these approaches is appropriate for criteria setting because tissue residues are the true measure of contaminant exposure. They integrate exposures over time, take into consideration bioavailability and assimilation efficiency of the compound, and incorporate biomagnification where it occurs.

Concentrations in aquatic organisms

Residue concentrations in organisms that are consumed by wildlife are measured and compared to dietary concentrations found to be toxic in controlled laboratory studies. Concentrations should be expressed on a dry weight basis both in test diets and in tissues of test organisms to correct for differing moisture contents across experimental diets and natural prey items.

When developing dietary toxicity thresholds, it must be remembered that the presence of the chemical could change the rate of food consumption, if a bad taste is imparted or an adverse reaction occurs that causes the animal to reduce consumption (see "Dietary Concern in the Development of WQC," p 96, for additional discussion regarding dietary issues). Thus, a reduction in food consumption might be the first sign of an adverse effect, and in fact, the actual dose ingested at high concentrations may be similar to that at lower concentrations. Varying consumption across dietary concentrations may affect risk calculations; alternatively, criteria might be based on dietary concentrations above which decreased food consumption affected weight or growth.

Concentrations in wildlife

Sampling wildlife species for residues is a valid means of measuring exposure for some contaminants. Currently, there is a large repository of literature describing concentrations of many bioaccumulative chemicals in tissues of wildlife species (Beyer et al. 1996). However, there is a severe limitation as to how to interpret these data because the relationship of residues to ecologically relevant effects is unknown in most cases. Furthermore, many chemicals accumulate in various tissues as a compensatory mechanism by the organism. Sequestration of chemicals in fat, feathers, bone, or proteins such as metallothionein may protect the organism from the effects of the toxicant unless the tissue reserve is mobilized. Therefore, increasing tissue concentrations will not always be correlated to increasing adverse effects.

Nonetheless, it is generally agreed that measurement of tissue residues, and relating these to adverse effects, is the most promising way to assess and to monitor potential effects of chemicals in organisms. The required information could be gathered as an adjunct to toxicity tests in the laboratory by conducting chemical analyses of tissues (e.g., fat, liver, kidney, brain, as appropriate for the particular chemical class). Concentrations in similar tissues in free-ranging wildlife could then be compared to the concentration–effects relationship and the probability of adverse effects estimated.

In the field, however, it can be difficult to define where the animal received its exposure. Migratory species, or those whose home ranges include several waterbodies, may contain residues representative of several sites. Thus, tissue residues, while helpful in monitoring and managing a wildlife species as a whole, may have less utility for establishing criteria for a specific body of water. Chemicals that accumulate and depurate quickly may be most amenable to residue-based criteria for specific sites. For example, selenium could be used as a site-specific indicator of exposure and effects because it is depurated by birds within about 2 weeks of cessation of exposure (Ohlendorf 1996).

Ingestion of sediment

Contaminated sediments are of concern to certain wildlife species such as waterfowl that eat rooted macrophytes, or shorebirds that ingest benthic invertebrates. These animals inadvertently ingest a large quantity of sediment, often as much as 15% of their total dietary intake (Beyer et al. 1994). If contaminant levels are high in the sediment but low in biota (as happens with metals such as lead, e.g.), the primary source of exposure might be from incidental sediment ingestion. However, not all of the contaminants in sediment are bioavailable via ingestion. Therefore, to include sediment exposure in criteria development, studies should be conducted with the representative taxa suggested below to determine estimated rates of sediment ingestion and bioavailability of contaminants of concern from ingested sediment. These 2 values then can be used to estimate the contribution that sediment ingestion makes towards the total critical dose.

Water ingestion

Certain avian wildlife species will consume as much as 30% of their body weight as water in a single day. However, this percentage may decrease to only a few percent, depending upon the water content of the food. Some species drink very little water at all. Regardless of their intake, our present knowledge suggests that if WQC for the protection of aquatic life are met, then consumption of water alone should not present a threat to wildlife species. Except for oil spills, dermal or inhalation exposures from water do not appear to be a threat to wildlife species and need not be considered. For amphibian eggs and larval stages, protection should be addressed through aquatic life criteria.

Migratory versus resident wildlife

Migratory and resident species present 2 exposure scenarios. Examples are Brant geese (*Branta bernicla*) that migrate through a region and reside in that region for a portion of a day or two, and the Canada goose (*Branta canadensis*) that may become resident and remain in an area year round. For contaminants that are present primarily in "hot spots," concerns for migratory species are for acute lethality because they are unlikely to remain in one place long enough to be affected by a low-level exposure. Resident species, on the other hand, are exposed over the longer term and should be protected from chronic effects on growth or reproduction as well. Migratory behavior actually encompasses a continuum of behaviors, rather than just these 2 discrete categories, and criteria development strategy must reflect this.

Selection of test species

Currently, the most pervasive difficulty in developing WC is insufficient data. Consistent with the requirements for testing aquatic biota, representatives of different wildlife groups should be tested to assess interspecific differences in sensitivity. For example, a database including 3 bird species, 3 mammal species, and a reptile could be the basis for criteria derivation (Table 3-1). Assessment of amphibians may overlap between aquatic life and WC because they represent aspects of both groups.

Currently, protocols exist for toxicity tests with mallards, mink, and deer mice. The other species recommended in Table 3-1 are amenable to captive rearing, with the exception of the northern short-tailed shrew, which has not been studied. Therefore, it is conceptually feasible to conduct toxicity tests across this range of species.

Test guidelines

Acute exposure involves determination of the LDx (lethal dose for x% population) value following a single exposure, whereas chronic effects could be based on

Table 3-1 Suggested wildlife taxa for criteria development

Representative group	Primary diet	Suggested species
Waterfowl	Rooted macrophytes and benthic invertebrates	Mallard (*Anas platyrhynchos*)
Shorebird	Benthic invertebrates	American avocet (*Recurvirostra americana*)
Songbird	Flying aquatic invertebrates	Redwinged blackbird (*Agelaius phoeniceus*)
Piscivorous mammal	Fish	Mink (*Mustela vison*)
Insectivorous mammal	Aquatic invertebrates	Northern short-tailed shrew (*Blarina brevicauda*)
Omnivorous mammal	Invertebrates and macrophytes	Deer mouse (*Peromyscus maniculatus*)
Reptile	Invertebrates and small aquatic vertebrates	Painted turtle (*Chrysemys picta*)

dietary exposures in either 28-day tests or longer reproduction studies, both of which assess sublethal responses in addition to lethality. The standard LD50 protocols for mallards, mink, and deer mice developed under the Federal Insecticide, Fungicide, and Rodenticide Act (FIFRA) and the Toxic Substances Control Act (TSCA), as modified by OECD, could be adopted in toto for aquatic criteria development (Menzer et al. 1994). Equivalent protocols should be developed for the additional recommended species. The avian reproduction test, as recommended by OECD, could be adopted for chronic testing of growth and reproduction endpoints for birds. A similar protocol developed under TSCA for small mammals could be used for the shrew and deer mouse. Protocols for acute and chronic mink tests also are available (Ringer et al. 1991). Songbird and reptile protocols will need to be developed de novo. Tissue residue analysis should be added to all of the existing and new test protocols. Finally, protocols for assessing bioavailability of contaminants in sediments also are needed to evaluate hazard from incidentally ingested sediment.

Standardizing test protocols is a critical need for WC. Most of the extant data on wildlife toxicology were not developed for purposes of establishing criteria and often lack key pieces of information (most notably, measurements of food consumption rates and tissue residue levels). However, where quality data exist (e.g., reproductive effects of PCBs in mink [Aulerich and Ringer 1977]), they should be used before generating new information. Increasing social emphasis on reducing the use of vertebrates in scientific testing may conflict with the technical need for additional toxicity data.

Recommendations

- Environmental criteria are needed to protect wildlife species from bioaccumulative compounds originating from aquatic ecosystems.
- WC should be based on chemical residues in tissues of target wildlife species, or in tissues of prey items.
- Sediment ingestion should be considered in deriving WC.
- Toxicity data to support WC should represent a range of wildlife species including bird, mammal, and reptilian species; assessment of amphibian species must be coordinated with aquatic life criteria.

Research needs

- Standardized test protocols are needed for several wildlife species, encompassing various taxonomic and ecological groups.
- Appropriate methods are needed for deriving criteria values from wildlife toxicity test results.

Dietary Concerns in the Development of WQC

There are 2 major concerns associated with the diet of aquatic organisms in the development of WQC. The first is the presence and potential effects of contaminants in the diet of aquatic organisms through direct toxicity interactions with essential nutrients, alteration of metabolism, appetite suppression, bioaccumulation, and biomagnification. The second is a data-quality issue associated with the interaction between nutrition and the relative sensitivity of organisms to toxicity. Inappropriate diet and media used in the culture of test organisms and during toxicity tests can result in additional stress, which may predispose test organisms to respond to lower toxicant concentrations than when cultured or tested under nutritionally suitable conditions. Conversely, organisms may develop tolerance to waterborne compounds by preexposure to the same or similar classes of toxicants in the diet.

Dietary toxicants

Current WQC consider dietary toxicants only in the context of establishing consumption guidelines for humans and some wildlife species to protect them from chemicals that biomagnify (e.g., PCBs, methylmercury). Diet also contributes directly to the toxicant body burden and toxicity in aquatic organisms in the absence of biomagnification. The contribution of dietary contaminants in fish, relative to waterborne contaminants, is usually important only with hydrophobic compounds when the concentration in the diet is about 5 orders of magnitude higher than in water (van Leeuwen and Hermens 1995). Many hydrophobic contaminants, such as DDT, PCBs, kepone, TCDD, and methylmercury, cause toxic effects when administered in the diet of fish (Buhler et al. 1969; Hawkes and Norris 1977; Mayer et al. 1977; Stehlik and Merriner 1983; Friedmann et al. 1996).

The importance of the dietary route of exposure on accumulation and toxicity is not limited to toxicants that biomagnify. Laboratory studies have shown that dietary metals and elements, such as selenium, can contribute significantly to body burdens in fish if elevated concentrations are present in the diet (Lanno et al. 1985; Coughlan and Velte 1989; Miller et al. 1993). Increased concentrations of dietary metals or selenium also have been linked with increasing body burdens in feral fish and may contribute to toxicity (Dallinger and Kautzky 1985; Coughlan and Velte 1989; Woodward et al. 1994, 1995).

Effects of dietary metals may vary between essential and nonessential metals due to homeostatic mechanisms that regulate the uptake, storage, and partitioning of essential metals. As a rule, waterborne metals are acutely toxic at lower concentrations than dietary metals, and metal absorbed from the diet is usually sequestered in a physiologically inert state (Lanno et al. 1987). A major unresolved question is the relative importance of dietary and waterborne routes of exposure in toxicity. Metals, especially essential metals, appear to partition differently within an organism depending upon whether they are taken up from the water or diet. Dietary metals are absorbed from the intestine, are passed into venous circulation, and are transported via the hepatic portal vein to the liver where they may be stored or transformed prior to distribution to other tissues. Waterborne metals initially bind to gill surfaces, and once absorbed, pass into the aortic system and are distributed to muscles and peripheral tissues prior to transport to the liver (Hodson and Hilton 1983).

Although dietary toxicants may contribute to the toxicant body burden, and even to the toxicity of that chemical to an organism, few protocols are available for dietary toxicity testing.

Nutritional considerations in toxicity testing

Water-quality criteria are derived from acute and chronic toxicity test data, but standardized toxicity test methods provide only simplified guidance on the nutrition of test organisms, both prior to and during the test (USEPA 1989). However, test organism nutrition modifies acute and chronic responses to waterborne toxicants by 2- to 10-fold (Holdway and Dixon 1985; Barry, Logan et al. 1995). Nutritional considerations are extremely important, especially during chronic toxicity tests where a major endpoint is growth, which is directly affected by food intake (Lanno et al. 1989).

Reduced food intake is a common response of test organisms exposed to water-borne toxicants (Lett et al. 1976; Kumaraguru and Beamish 1986; Lanno and Dixon 1996) but is rarely monitored during toxicity tests. Food intake can be monitored relatively easily in studies using larger fish but is more difficult with larval fish and invertebrates. In chronic toxicity tests, such as the 7-day larval fathead minnow growth and survival test and the *Ceriodaphnia* sp. reproduction test, food refusal becomes extremely important due to the higher specific metabolic rate and limited energy reserves of these smaller organisms. Energetic limitations due to reduced food intake, rather than direct effects of toxicants on metabolism, may be the underlying cause of observed effects such as reduced growth or reproduction. Current toxicity test methods do not provide guidance on addressing effects due to reduced food intake.

Diet as a confounding factor in toxicity tests

The metabolic changes that result from the presence of food in the gut, as well as the presence of dietary toxicants, may alter physiological responses of test organisms to waterborne toxicants. Ingestion of food sets off a cascade of events that increases the metabolic rate of test organisms (Jobling 1981) and that alters the uptake, metabolism, and excretion of both dietary and waterborne contaminants (e.g., Jimenez et al. 1987; van Veld et al. 1988). Absence of food from the gastrointestinal tract also may alter other facets of metabolism (e.g., Andersson et al. 1985; Boon and Duinker 1985; Watt et al. 1988). Normal physiological parameters, such as plasma levels of thyroid hormones, glucose, cortisol, and lipid, change in response to relatively short periods of fasting or food deprivation (48 to 72 hours) (Eales et al. 1981; White and Fletcher 1986). Most of these observations are based on tests conducted with relatively large fish; for smaller test organisms with higher specific metabolic rates, feeding becomes a much more important issue. For smaller test organisms in chronic tests, food refusal may result in the complete alteration of the physiological state of the organism during the course of the test and in dramatically different toxicity.

If the diet contains toxicants, effects of feeding may be greater than simply altering metabolic rate. The intestinal metabolism of B(*a*)P by toadfish (*Opsanus tau*) preexposed to dietary B(*a*)P was greater (90%) than in fish preexposed to a control diet containing no B(*a*)P (60%) (van Veld et al. 1988). Dietary metals may also have an effect on subsequent metal exposure. Preexposure of trout to elevated concentrations of dietary copper increased their tolerance of waterborne copper (Miller et al. 1993).

Feeding of test organisms during acute toxicity tests is usually discouraged because of concerns over the effects of food particles, DOC, and fecal matter altering chemical bioavailability. However, feeding is necessary during chronic tests to avoid starvation (Sprague 1973; USEPA 1989). During chronic tests, food is usually administered to the test chamber at a fixed rate (e.g., a specified volume of an algal or *Artemia* suspension) for the duration of the test, with no consideration of the effects that growth or mortality may have on food intake. The question arises as to whether suggested feeding regimes are realistic in reproducing the natural feeding regimes of test organisms or whether this should even be a concern. The natural feeding pattern of test organisms (e.g., single large meals, continuous foraging) is not usually considered, and one of a number of arbitrary or convenient feeding regimes is imposed, usually ad libitum feeding once or twice per day. With larger test organisms, it is possible to feed to satiety at each feeding and to quantify food intake, but with smaller organisms, excess food is offered and uneaten food may be removed at some time during the test. Larger organisms may also be fed at a rate relative to the weight of the test organism. Feeding tables that are based upon size of fish and water temperature and that allow specific growth rates to be achieved (Hilton and Slinger 1981) are available for trout, but little guidance is available for other standard test species. However, neither of these approaches allows the separation of effects due to toxicant exposure from effects due to reduced food intake. Experimental designs incorporating "pair feeding" can distinguish between responses due directly to the toxicant and those due to differences in food intake (Baker 1986; Lanno and Dixon 1996). Each treatment replicate is fed, the amount of food consumed is monitored, and a randomly matched control replicate receives the same amount of food given to the treatment replicate. If decreased food intake is an effect of toxicant exposure, then the ration given to the matched control is reduced accordingly. Toxicant exposure is thus normalized to food intake, allowing differences in endpoints to be attributed directly to the level of toxicant rather than to the reduced food intake. Further complications arise when the diet is used as a route of exposure.

Another issue in the nutrition of test organisms is the presence or absence of essential nutrients in test and culture media. A deficiency of essential nutrients may compromise the health of test organisms and confound their responses during toxicity tests. Current laboratory toxicity tests often expose organisms in reconstituted media comprised of deionized water with added inorganic salts, ensuring a specific hardness, alkalinity, and pH (USEPA 1989). The preparation of reconstituted water minimizes the chance of introducing toxicants other than the chemical or element to be tested, and exposure of the test organism to the toxicant of interest is maximized. As a result, essential elements such as copper and zinc may be lacking in the test medium because these are not added in the preparation of the reconstituted water.

If culture media are deficient in essential elements such as copper, test organisms may be at risk for copper deficiency at the beginning of the test (e.g., see discussion regarding algal nutrition in "Effects on plants," p 86). Traditionally, culture media used in standardized protocols for most aquatic test organisms are designed to keep metal concentrations as low as possible to avoid toxicity. In fact, it may be necessary to supplement culture media with essential nutrients such as selenium and vitamin B_{12} to avoid nutrient deficiency conditions (Keating and Dagbuson 1984; Keating 1985).

Test organisms can acclimate to a fairly wide range of essential element concentrations before entering a zone of toxicity (van Tilborg and van Assche 1998). Because acclimation to metals prior to exposure during a toxicity test can influence test results (Dixon and Sprague 1981), special attention should be given to the levels of metals in culture and test media. Test organisms adapted to extremely low concentrations of metals prior to testing may be much more sensitive during subsequent metal exposures than organisms acclimated to higher concentrations. Further research is needed to define the relationship between gastrointestinal and gill membrane uptake of metals and other essential elements in satisfying nutritional requirements.

A final consideration is the baseline concentration of toxicants that occur in diets fed to test organisms prior to and during toxicity tests. Commercially available food organisms and formulated test diets are far from uncontaminated, and vary considerably in their baseline or background concentrations of contaminants. Cech and Choi (1998) have reported on elevated mercury in commercial fish feeds. Bengtson et al. (1985) outlined the geographical variation in organic xenobiotic and heavy metal content in *Artemia nauplii*. Commercially prepared diets (e.g., trout chow, flake fish food) will vary in baseline toxicant concentrations due to variable sources of ingredients (e.g., fish meal) or production technologies (e.g., rolling between metal-plated rollers). As a routine procedure, concentrations of contaminants should be measured and reported for all test organism diets used in cultures or toxicity tests (USEPA 1989).

Recommendations

- Diet type, source, and feeding regime used during toxicity tests should be stated.
- Background toxicant concentrations in diets should be measured and reported.
- If toxicant exposure reduces growth, food intake should be monitored; pair-feeding experimental designs may be used to delineate the role of reduced food intake in reducing growth.
- Culture and test media should contain essential elements at concentrations that meet nutritional requirements and approximate a mean or median

background concentration for the specific element in the region where test results will be applied.

Population, Community, and Ecosystem Responses

Ecological significance and level of biological organization

Although standard acute and chronic toxicity tests form the foundation for the development of WQC, other types of information should be integrated into this process (Table 3-2). Because effects of contaminants on aquatic organisms may be manifested at all levels of biological organization (Clements and Kiffney 1994), the suite of endpoints employed to establish WQC also should include community and ecosystem responses (Figure 3-7). Proponents of reductionist approaches assert that effects of contaminants on ecosystem structure and function can be predicted from simple laboratory bioassays. However, a holistic perspective recognizes that

Table 3-2 Advantages, limitations, and examples of biomonitoring and experimental approaches for assessing effects of chemicals on ecologically significant endpoints

Approach	Advantages	Limitations	Examples
Routine biomonitoring	Large existing database Simple design	Confounding variables Pseudoreplication Highly site specific No cause or effect	Clements 1994
Contamination gradient	Moderate to strong cause or effect	Availability of suitable sites Presence of confounding factors	Quinn et al. 1992 Hickey et al. 1995 Hickey and Clements 1998
Spatially extensive biomonitoring	Large spatial scale Broad generalizations	Large data requirements No cause or effect	Feldman and Connor 1992 Clements and Kiffney 1995
Before-after control-impact (BACI)	Some cause or effect Statistically rigorous	Lack of pre-impact data	Stewart-Oaten et al. 1986 Wiens and Parker 1995
Ecosystem manipulations	Moderate cause or effect Includes top predators	Limited replication Logistically difficult Costly	Rapport et al. 1985 Schindler 1987
Microcosms and mesocosms	Strong cause or effect Can examine chemical interactions	Small spatial, temporal scale May lack realism	Odum 1984 Kiffney and Clements 1996

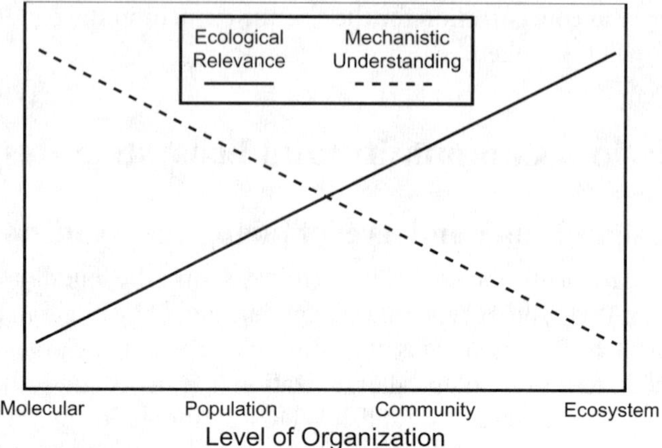

Figure 3-7 Hypothetical relationship between levels of biological organization, mechanistic understanding, and ecological relevance

some emergent properties of ecosystems cannot be predicted from their component parts and that single species cannot fully represent the behaviors of communities and ecosystems. Schindler et al. (1985) reported that loss of lake trout in acidified lakes was an indirect effect of a depleted prey base and not a result of direct toxicity. Similarly, Gonzalez and Frost (1994) found major discrepancies in effects of acidification on rotifers between results of laboratory bioassays and a whole lake experiment.

Responses at lower levels of organization, such as those included in standard toxicity tests, are linked directly to exposure, occur faster, and can be interpreted mechanistically. In contrast, responses at higher levels of organization are more complex, often lack a simple mechanistic explanation, but are more ecologically relevant. Establishing causal linkages across levels of biological organization will help verify the ecological relevance of molecular and biochemical responses. Understanding the molecular and biochemical mechanisms responsible for changes in populations and communities will improve our ability to link observed responses directly to specific contaminants.

Assessments of ecological effects in the field generally involve measurement of structural and functional endpoints. Typical structural measures of chemical effects include reduced abundance, reduced species richness, and shift in community composition from sensitive to tolerant species. More sophisticated pollution and multi-metric indices have been derived from these data, based on the assumption that a shift in community composition at polluted sites relative to unpolluted sites is an indication of the degree of contamination (Karr 1991). Because these assessments often involve analysis of numerous biotic and abiotic variables, multivariate approaches have been proposed to examine relationships between species assemblages and multiple environmental variables. Multivariate approaches reduce complex, multidimensional data to 2 or 3 dimensions, and allow researchers to identify key environmental variables responsible for patterns of species abundance.

Functional measures of ecological integrity generally include endpoints such as primary and secondary productivity, nutrient cycling, energy flow, or detritus processing. As with community-level endpoints, the usefulness of these measures for assessing effects of chemicals will probably vary among ecosystems. Rapport et al. (1985) described an "ecosystem distress syndrome" and identified general ecosystem responses to several types of disturbance, including loss of nutrients, decreased primary productivity, reduced species diversity, reduction in the size of organisms, and shifts in community composition.

Approaches for investigating ecologically significant responses

Because most field investigations are descriptive and often restricted to a single system (Clements 1994), the inferences from these studies are relatively weak and restricted to the specific system being investigated. Results of field studies also are confounded by numerous biotic and abiotic factors, and therefore establishing direct cause-and-effect relationships between chemicals and biological responses is often not possible. The inability to replicate or randomize treatments among locations is recognized as the most serious statistical problem of field studies. Support for a direct causal relationship between specific chemicals and responses is strengthened by more sophisticated sampling designs (e.g., before–after control–impact [BACI] designs), weight-of-evidence approaches, contamination gradient studies, and the use of experimental procedures (e.g., microcosms, mesocosms, and field manipulations). Wiens and Parker (1995) provide an excellent overview of various approaches used to assess environmental impacts based on their experiences with the *Exxon Valdez* oil spill. These authors also provide guidance on the relative severity of pseudoreplication in various study designs, as well as recommendations to reduce these effects.

Contamination gradients

Field studies that incorporate concentration–response studies along a contamination gradient provide an opportunity for deriving regression relationships for responses at the species and community levels. The contamination gradient approach has been used to investigate body burden and physiological response relationships for bivalve mollusks (e.g., Hickey et al. 1995) and is widely used in sediment studies. Studies in rivers with comparable contaminant sources but differing levels of available dilution have been used to develop response relationships for macroinvertebrate communities exposed to inorganic suspended sediments (Quinn et al. 1992), heavy metals (Hickey and Clements 1998), and complex mixtures such as waste stabilization lagoon effluent (Quinn and Hickey 1993). Determining cause–effect relationships for complex mixtures may be difficult, given that multiple contaminants and combined stressors may be contributing to the observed effects, but useful thresholds may be developed that have, at a minimum, site-specific applicability.

Spatially extensive sampling designs

The development of more sophisticated sampling designs that employ multiple reference and contaminated sites also has been proposed as an alternative to traditional field approaches used in most biomonitoring studies (Feldman and Connor 1992; Clements and Kiffney 1995; Resh et al. 1995). Although such spatially extensive assessments often lack estimates of temporal variation, they can be used to characterize ecological conditions at a regional scale. In addition, spatially extensive monitoring within an ecoregion provides the necessary background information that can be employed to evaluate future changes in water quality (USEPA 1995a).

BACI designs

An alternative to the traditional field approach for assessing causality is the use of the BACI design (Stewart-Oaten et al. 1986). This powerful experimental design involves the use of both spatial and temporal treatments. The magnitude of difference between control and impact sites is measured before and after impact occurs. The BACI design is recognized as one of the best models for impact assessment and has been employed to estimate effects of a variety of chemical stressors on aquatic ecosystems. The most serious limitation of this approach is that pre-impact data are rarely available.

Whole ecosystem manipulations

Certainly the most direct way to investigate the effects of chemicals in the field is by direct ecosystem manipulation. Although ecologists have long recognized the usefulness of field manipulations for determining cause and effect relationships, these approaches have received limited attention in ecotoxicology. Important exceptions are the whole lake acidification studies conducted in the Experimental Lakes Area, Canada and Little Rock Lake, Wisconsin, USA (Schindler 1987; Brezonik et al. 1993), pesticide manipulations conducted in streams in the Coweeta Experimental Forest, USA (Wallace et al. 1986), and copper studies at Shaylor Run (Winner et al. 1980). One of the more consistent results of these studies is the finding that indirect effects are often more significant than direct effects. For example, whole lake acidification studies by Schindler (1987) demonstrated that reduced prey abundance had a greater effect on lake trout populations than direct toxicological effects of reduced pH.

Microcosms and mesocosms

Whole ecosystem manipulations are limited because of the expense, logistical difficulties, and legal or ethical issues associated with the intentional introduction of chemicals into natural ecosystems. The use of microcosms and mesocosms is an important compromise between simple field assessments and logistically difficult and impractical field manipulations. Mesocosms simulating stream ecosystems have been widely used and differ greatly in size and exposure and duration (Gluckert 1993). Experimental designs are flexible, ranging from replicated treatments (analysis of variance [ANOVA]) to concentration–response designs (e.g., Liber et al. 1992). These systems can be used to isolate effects of specific environmental variables on ecosystem structure and function. In addition, they can be employed to examine interactions among stressors, or between stressors and other abiotic factors. Finally, because microcosms and mesocosms generally employ multispecies assemblages, they provide an efficient approach to generate sensitivity data for a large number of indigenous species. The emergence of microcosms and mesocosms as a tool to assess effects of chemicals represents an important supplement to single-species approaches (Odum 1984).

Recovery of aquatic ecosystems from chemical stress

Natural variation in biotic (species life-history characteristics, community composition) and abiotic (geomorphology, hydrology) characteristics of aquatic ecosystems will influence the magnitude of the response to chemical stressors and the time required to recover (i.e., respond to the removal of chemical stress; Figure 3-8). If some ecosystems are inherently more fragile than others, identifying characteristics that increase the amount of time required for ecosystem recovery is an important topic of research. Factors such as temporal variability, habitat

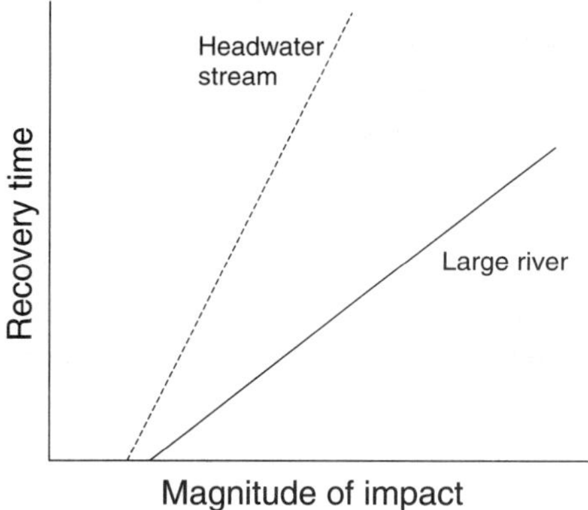

Figure 3-8 Hypothetical recovery time of headwater streams and large rivers as a function of magnitude of disturbance

heterogeneity, and magnitude of the perturbation are expected to influence ecosystem resilience (Poff and Ward 1990). For example, it is hypothesized that systems with significant potential for recolonization (e.g., streams with large, unpolluted tributaries) will recover faster than small, isolated systems. Streams characterized by variable hydrologic regimes that are dominated by opportunistic species (Poff and Ward 1990; Palmer 1995) also would show greater resilience than stable streams dominated by long-lived species. Howarth (1991) speculated that ecosystems with fewer opportunistic species, lower diversity, and closed element cycles would be sensitive to contaminants. Using microcosm experiments and field assessments, Kiffney and Clements (1996) showed that effects of heavy metals were greater on benthic communities from small headwater streams compared to similar communities from larger rivers. Although it is currently not possible to predict the precise recovery time for specific ecosystems, these issues should be considered when establishing site-specific WQC.

Recommendation

- Develop approaches to integrate field, microcosm, and mesocosm data into the establishment of WQC.

Research needs

- Develop an understanding of the biotic and abiotic factors that influence recovery of ecosystems from chemical stress.
- Develop scientifically credible methods for incorporating population and community-level endpoints into criteria development.

References

Aldenberg T, Slob W. 1993. Confidence limits for hazardous concentrations based on logistically distributed NOEC toxicity data. *Ecotox Environ Saf* 25:48-63.

Amato JR, Mount DI, Durhan EJ, Lukasewycz MT, Ankley GT, Robert ED. 1992. An example of the identification of diazinon as a primary toxicant in an effluent. *Environ Toxicol Chem* 11:209-216.

Andersson T, Koivusaari U, Forlin L. 1985. Xenobiotic biotransformation in the rainbow trout liver and kidney during starvation. *Comp Biochem Physiol* 82C:221-225.

Ankley G, Bradbury S, Hermens J, Mekenyan O, Tollefson K-E. 1997. Current approaches to the use of structure activity relationships (SARs) in identifying the hazards of endocrine disrupting chemicals to wildlife. In: Tattersfield L, Matthiessen P, Campbell P, Grandy N, Lange R, editors. Workshop on Endocrine Modulators, Wildlife-Assessment and Testing. Brussels, Belgium: SETAC. p 19-40.

Ankley GT, Dierkes JR, Jensen DA, Peterson GS. 1991. Piperonyl butoxide as a tool in aquatic toxicological research with organophosphate insecticides. *Ecotoxicol Environ Saf* 21:266-274.

Ankley GT, Giesy JP. 1998. Endocrine disruptors in wildlife: A weight of evidence perspective. In: Kendall R, Dickerson R, Suk W, Giesy JP, editors. Principles and processes for assessing endocrine disruption in wildlife. Pensacola FL, USA. Society of Environmental Toxicology and Chemistry (SETAC). p 349-368.

Ankley GT, Johnson RD, Detenbeck NE, Bradbury SP, Toth G, Folmar LC. 1997. Development of a research strategy for assessing the ecological risk of endocrine disruptors. *Rev Toxicol* 1:71-106.

Ankley GT, Schubauer-Berigan MK, Monson PD. 1995. Influence of pH and hardness on toxicity of ammonia to the amphipod *Hyalella azteca*. *Can J Fish Aquat Sci* 52:2078-2083.

Arthur AD. 1991. Verification studies of a body residue based model for predicting the sublethal toxicity of time variable exposures to organic chemicals in small fish [MSc Thesis]. Waterloo ON, Canada: University of Waterloo, Department of Biology.

[ASTM] American Society for Testing and Materials. 1991. Standard guide for conducting static toxicity tests with *Lemna gibba*. In: Annual book of standards. Volume 11.05. West Conshohocken PA, USA: ASTM. G3,E1415 91.

[ASTM] American Society for Testing and Materials. 1998. Standard guide for conducting static 96 hour toxicity tests with microalgae. In: Annual book of standards. Volume 11.05. West Conshohocken PA, USA: ASTM. E1218 97a.

Aulerich RJ, Ringer RK. 1977. Current status of PCB toxicity to mink, and effects on their reproduction. *Arch Environ Contam Toxicol* 8:487-498.

Bailey HC, DiGiorgio C, Kroll K, Miller JL, Hinton DE, Starrett G. 1996. Development of procedures for identifying pesticide toxicity in ambient waters: Carbofuran, diazinon, chlorpyrifos. *Environ Toxicol Chem* 15:837-845.

Baker DH. 1986. Problems and pitfalls in animal experiments designed to establish dietary requirements for essential nutrients. *J Nutr* 116:2339-2349.

Barry MJ, Logan DC, Ahokas JT, Holdway DA. 1995. Effect of algal food concentration on toxicity of two agricultural pesticides to *Daphnia carinata*. *Ecotox Environ Saf* 32:273-279.

Barry MJ, O'Halloran K, Logan DC, Ahokas JT, Holdway DA. 1995. Sublethal effects of esfenvalerate pulse exposure on spawning and non-spawning Australian crimson spotted rainbowfish (*Melanotaenia fluviatilis*). *Arch Environ Contam Toxicol* 28:459-463.

Bellrose FC. 1976. Ducks, geese, and swans of North America. 2nd edition. Harrisburg PA, USA: Stackpole Books. 354 p.

Benenati F. 1990. Keynote address: Plants keystone to risk assessment. In: Wang W, Gorsuch J, Lower L, editors. Plants for toxicity assessment. Philadelphia PA, USA: ASTM. STP 1091. p 5-13.

Bengtson DA, Beck AD, Simpson KL. 1985. Standardization of the nutrition of fish in aquatic toxicological testing. In: Cowey CB, Mackie AM, Bell JB, editors. Nutrition and feeding in fish. London, UK: Academic Press. p 431-446.

Bennett RS, Schafer DW. 1988. Procedures for evaluating the potential of birds to avoid chemically-contaminated food. *Environ Toxicol Chem* 7:359-362.

Benoit DA, Sibley PK, Juenemann JL, Ankley GT. 1997. *Chironomus tentans* life cycle test: Design and evaluation for use in assessing toxicity of contaminated sediments. *Environ Toxicol Chem* 16:1165-1176.

Bergman HL, Dorward-King EJ, editors. 1997. Reassessment of metals criteria for aquatic life protection: Priorities for research and implementation. SETAC Pellston Workshop on Reassessment of Metals Criteria for Aquatic Life Protection; 1996 Feb 10-14; Pensacola FL, USA. Society of Environmental Toxicology and Chemistry (SETAC). 114 p.

Beyer WN, Connor EE, Gerould S. 1994. Estimates of soil ingestion by wildlife. *J Wildl Manag* 58:375-382.

Beyer WN, Heinz GH, Redmon-Norwood AW. 1996. Environmental contaminants in wildlife: Interpreting tissue concentrations. Boca Raton FL, USA: Lewis Publishers.

Boon JP, Duinker JC. 1985. Kinetics of polychlorinated biphenyl (PCB) components in juvenile sole (*Solea*) in relation to concentrations in water and to lipid metabolism under conditions of starvation. *Aquat Toxicol* 7:119-134.

Borgmann U, Borgmann AI. 1997. Control of ammonia toxicity to *Hyalella azteca* by sodium, potassium and pH. *Environ Pollut* 95:325-331

Bowers WS. 1990. Prospects for the use of insect growth regulators in agriculture. In: Hoshi M, Yamashita O, editors. Advances in invertebrate reproduction 5. New York NY, USA: Elsevier. p 365-382.

Bradbury SP. 1994. Predicting modes of toxic action from chemical structure: An overview. *SAR QSAR Environ Res* 2:89-104.

Bradbury SP. 1995. Quantitative structure activity relationships and ecological risk assessment: An overview of predictive aquatic toxicology research. *Toxicol Lett* 79:229-237.

Bradbury SP. 1996. 2,3,7,8-tetrachlorodibeno-p-dioxin. In: Fairbrother A, Locke LN, Hoff GL, editors. Noninfectious diseases of wildlife. 2nd ed. Ames IA, USA: Iowa State University Press. p 87-98.

Breck JE. 1988. Relationships among models for acute toxic effects: Applications to fluctuating concentrations. *Environ Toxicol Chem* 7:775-778.

Brezonik PL, Eaton JG, Frost TM, Garrison PJ, Kratz TK, Mach CE, McCormick JH, Perry JA, Rose WA, Sampson CJ, Shelley BCL, Swenson WA, Webster KE. 1993. Experimental acidification of Little Rock Lake, Wisconsin: Chemical and biological changes over the pH range 6.1 to 4.7. *Can J Fish Aquat Sci* 50:1101-1121.

Broderius SJ. 1991. Modeling the joint toxicity of xenobiotics to aquatic organisms: Basic concepts and approaches. In: Aquatic toxicology and risk assessment. 14th Volume. Philadelphia PA, USA: ASTM. p 107-127.

Broderius SJ, Kahl MD. 1985. Acute toxicity of organic chemical mixtures to the fathead minnow. *Aquat Toxicol* 6:307-322.

Buhler DR, Rasmusson ME, Shanks WE. 1969. Chronic oral DDT toxicity in juvenile coho and chinook salmon. *Toxicol Appl Pharmacol* 14:535-555.

Burgess RM, McKinnery RA, Brown WA, Quinn JG. 1996. Isolation of marine sediment colloids and associated polychlorinated biphenyls: An evaluation of ultrafiltration and reverse phase chromatography. *Environ Sci Technol* 30:1923-1932.

Carlson AR, Brungs WA, Chapman GA, Hansen DJ. 1984. Guidelines for deriving numeral aquatic site-specific water quality criteria by modifying national criteria. Duluth MN, USA: USEPA. EPA-600-3-84-099.

[CCREM] Canadian Council of Resource and Environment Ministers. 1987. Canadian water quality guidelines. Ottawa ON, Canada: CCREM, Task Force on Water Quality Guidelines.

Cech JJ, Choi MH. 1998. Unexpectedly high mercury levels in pelleted commercial fish feed. *Environ Toxicol Chem* 17:1979-1981.

Chew RD, Hamilton MA. 1985. Toxicity curve estimation: Fitting a compartment model to median survival times. *Trans Am Fish Soc* 114:403-412.

Clements WH. 1994. Benthic invertebrate community responses to heavy metals in the Upper Arkansas River Basin, Colorado. *J North Am Benthol Soc* 13:30-44.

Clements WH, Kiffney PM. 1994. Assessing contaminant impacts at higher levels of biological organization. *Environ Toxicol Chem* 13:357-359.

Clements WH, Kiffney PM. 1995. The influence of elevation on benthic community responses to heavy metals in Rocky Mountain streams. *Can J Fish Aquat Sci* 52:1966-1977.

Connell DW. 1988. Bioaccumulation behaviour of persistent organic chemicals with aquatic organisms. *Rev Environ Contam Toxicol* 101:117-154.

Connolly JP. 1985. Predicting single-species toxicity in natural water systems. *Environ Toxicol Chem* 4:573-582.

Cook PM, Erickson RJ, Spehar RL, Bradbury SP, Ankley GT. 1993. Interim report on data and methods for assessment of 2,3,7,8-tetrachlorodibenzo-p-dioxin risks to aquatic life and associated wildlife. Duluth MN, USA: USEPA. EPA-600-R-93-055.

Cook PM, Zabel EW, Peterson RE. 1997. The TCDD toxicity equivalence approach for characterizing risks for early life-stage mortality in trout. In: Rolland RM, Gilbertson M, Peterson RE, editors. Chemically induced alterations in functional development and reproduction of fishes. Pensacola FL, USA. Society of Environmental Toxicology and Chemistry (SETAC). p 9-28.

Cooper RL, Stoker TE, Tyrey L, Goldman JM, McElroy WK. 2000. Atrazine disrupts the hypothalamic control of pituitary-ovarian function. *Toxicol Sci* 53:297-307.

Corn PS. 2000. Amphibian declines: Review of some current hypotheses. In: Sparling DW, Bishop CA, Linder G, editors. Ecotoxicology of amphibians and reptiles. Pensacola FL, USA: SETAC. 904 p.

Coughlan DJ, Velte JS. 1989. Dietary toxicity of selenium-contaminated red shiners to striped bass. *Trans Am Fish Soc* 118:400-408.

Dallinger R, Kautzky H. 1985. The importance of contaminated food for the uptake of heavy metals by rainbow trout (*Salmo gairdneri*): A field study. *Oecologia* 67:82-89.

De Groot MJ, van der Aar EM, Nieuwenhuizen PJ, van der Plas RM, den Kelder GMD, Commandeur JNM, Vermeulen NPE. 1995. A predictive substrate model for rat glutathione s-transferase 4-4. *Chem Res Toxicol* 8:649-658.

Delos CG. 1994. Possible revisions to EPA's procedure for deriving aquatic life criteria. Water Environment Federation Annual Meeting; 1994 Oct 17-19; Chicago IL, USA.

Dixon DG, Sprague JB. 1981. Acclimatization to copper by rainbow trout (*Salmo gairdneri*)—A modifying factor in toxicity. *Can J Fish Aquat Sci* 38:880-888.

Donohoe RM, Curtis LR. 1996. Estrogenic activity of chlordecone, o,p'-DDE in juvenile rainbow trout: Induction of vitellogenesis and interaction with hepatic estrogen binding sites. *Aquat Toxicol* 36:31-52.

Eales JG, Hughes M, Uin L. 1981. Effect of food intake on diel variation in plasma thyroid hormone levels in rainbow trout, *Salmo gairdneri*. *Gen Comp Endocrinol* 45:167-174.

Elonen GE, Spehar RL, Holcombe GW, Johnson RD, Fernandez JD, Erickson RJ, Tietge JE, Cook PM. 1998. Comparative toxicity of 2,3,7,8-tetrachlorodibenzo-p-dioxin to seven freshwater fish species during early life-stage development. *Environ Toxicol Chem* 17:472-483.

Erickson RJ. 1985. An evaluation of mathematical models for the effects of pH and temperature on ammonia toxicity to aquatic organisms. *Water Res* 19:1047-1058.

Erickson RJ, Ankley GT, DeFoe DL, Kosian PA, Makynen EA. 1999. Additive toxicity of binary mixtures of phototoxic polycyclic aromatic hydrocarbons to the oligochaete *Lumbriculus variegatus*. *Toxicol Appl Pharmacol* 154:97-105.

Erickson RJ, Benoit DA, Mattson VR. 1987. A prototype toxicity factors model for site-specific copper water quality criteria. Duluth MN, USA: USEPA.

Erickson RJ, Kleiner C, Fiandt J, Highland T. 1989. Report on the feasibility of predicting the effects of fluctuating exposures and application to water quality criteria. Duluth MN, USA: USEPA. EPA-600-X-89-307.

Fairchild JF, La Point TW, Zajicek JL, Nelson MK, Dwyer FJ, Lovely PA. 1992. Population, community, and ecosystem level responses of aquatic mesocosms to pulsed doses of a pyrethroid insecticide. *Environ Toxicol Chem* 11:115-129.

Feldman RS, Connor EF. 1992. The relationship between pH and community structure of invertebrates in streams of the Shenandoah National Park, Virginia, USA. *Freshw Biol* 27:261-276.

Friedmann AS, Watzin MC, Brinck-Johnsen T, Leiter JC. 1996. Low levels of dietary methylmercury inhibit growth and gonadal development in juvenile walleye (*Stizostedion vitreum*). *Aquat Toxicol* 35:265-278.

Gagen CJ, Sharpe WE, Carline RF. 1993. Mortality of brook trout, mottled sculpins, and slimy sculpins during acidic episodes. *Trans Am Fish Soc* 122:616-628.

Gluckert JB. 1993. Artificial streams in ecotoxicology. *J North Am Benthol Soc* 12:313-384.

Gobas FAPC, Zhang X. 1992. Measuring bioconcentration factors and rate constants of chemicals in aquatic organisms under conditions of variable water concentrations and short exposure time. *Chemosphere* 25:1961-1971.

Gobas FAPC, Zhang X. 1994. Interactions of organic chemicals with particulate and dissolved organic matter in the aquatic environment. In: Hamelink JL, Landrum PF, Berman HL, Benson WH, editors. Bioavailability: Physical, chemical and biological interactions. Boca Raton FL, USA: Lewis Publishers. p 83-91.

Gonzalez MJ, Frost TM. 1994. Comparisons of laboratory bioassays and a whole lake experiment: Rotifer responses to experimental acidification. *Ecol Appl* 4:69-80.

Grothe DR, Dickson KL, Reed-Judkins DK, editors. 1996. Whole effluent toxicity testing: An evaluation of methods and prediction of receiving system impacts. Pensacola FL, USA: SETAC. 346 p.

Gupta UC, Jame YW, Campbell CA, Leyshon AJ, Nicholaichuk W. 1985. Boron toxicity and deficiency: A review. *Can J Soil Sci* 65:381-409.

Handy RD. 1992a. The assessment of episodic metal pollution. 1. Uses and limitations of tissue contaminant analysis in rainbow trout (*Oncorhynchus mykiss*) after short waterborne exposure to cadmium or copper. *Arch Environ Contam Toxicol* 22:74-81.

Handy RD. 1992b. The assessment of episodic metal pollution. 2. The effects of cadmium and copper enriched diets on tissue contaminant analysis in rainbow trout (*Oncorhynchus mykiss*). *Arch Environ Contam Toxicol* 22:82-87.

Hawkes CL, Norris LA. 1977. Chronic oral toxicity of 2,3,7,8-tetrachlorodibenzo-*p*-dioxin (TCDD) to rainbow trout. *Trans Am Fish Soc* 106:641-645.

Heath AG. 1995. Water pollution and fish physiology. 2nd ed. Boca Raton FL, USA: Lewis Publishers. 359 p.

Heming TA, Sharma A, Kumar Y. 1989. Time toxicity relationships in fish exposed to the organochlorine pesticide methoxychlor. *Environ Toxicol Chem* 8:923-932.

Herman RL, Kincaid HL. 1988. Pathological effects of orally administered estradiol to rainbow trout. *Aquaculture* 72:165-172.

Hermens J, Canton H, Steyger N, Wegman R. 1984. Joint effects of a mixture of 14 chemicals on mortality and inhibition of reproduction of *Daphnia magna*. *Aquat Toxicol* 5:315-322.

Hickey CW, Blaise C, Constan G. 1991. Microtesting appraisal of ATP and cell recovery toxicity end points after acute exposure of *Selenastrum capricornutum* to selected chemicals. *Environ Toxicol Water Qual* 6:383-403.

Hickey CW, Clements WH. 1998. Effects of heavy metals on benthic invertebrate communities in New Zealand streams. *Environ Toxicol Chem* 17:2338-2346.

Hickey CW, Roper DS, Holland PT, Trower TM. 1995. Accumulation of organic contaminants in two sediment dwelling shellfish with contrasting feeding modes: Deposit- (*Macoma liliana*) and filter-feeding (*Austrovenus stutchbury*i). *Arch Environ Contam Toxicol* 29:221-231.

Hickie BE. 1990. Development and testing of a toxicokinetic based model for pulse exposure toxicity of organic contaminants [DPhil Thesis]. Waterloo ON, Canada: University of Waterloo. 109 p.

Hilton JW, Slinger SJ. 1981. Nutrition and feeding of rainbow trout. *Can Spec Publ Fish Aquat Sci* 55:15.

Hodson PV, Blunt BR, Borgmann U, Minns CK, McGaw SM. 1983. The effect of fluctuating lead exposures on lead accumulation by rainbow trout (*Salmo gairdneri*). *Environ Toxicol Chem* 2:225-239.

Hodson PV, Hilton JW. 1983. The nutritional requirements and toxicity to fish of dietary and waterborne selenium. In: Hallberg R, editor. Environmental biogeochemistry. *Ecol Bull* 35:335-340.

Holdway DA, Barry MJ, Logan DC, Robertson D, Young V, Ahokas JT. 1994. Toxicity of pulse exposed fenvalerate and esfenvalerate to larval Australian crimson spotted rainbow fish (*Melanotaenia fluviatilis*). *Aquat Toxicol* 28:169-187.

Holdway DA, Dixon DG. 1985. Acute toxicity of pulse-dosed methoxychlor to juvenile flagfish (*Jordanella floridae*) as modified by age and food availability. *Aquat Toxicol* 6:243-250.

Hollebone BR, Brownlee LJ, Davis D, Michelin N, Purdy D. 1995. A mechanistic structure-activity relationship for hepatic polysubstrate monoxygenase. *Environ Toxicol Chem* 14:29-41.

Hollis L, McGeer JC, McDonald DG, Wood CM. 1999. Cadmium accumulation, gill Cd binding, acclimation, and physiological effects during long term sublethal Cd exposure in rainbow trout. *Aquat Toxicol* 46:101-119

Holst RW, Ellwanger TC. 1982. Pesticide assessment guidelines, subdivision J hazard evaluation: Non-target plants. Washington DC, USA: USEPA. EPA-540-9-82-020.

Howarth RW. 1991. Comparative responses of aquatic ecosystems to toxic chemical stress. In: Cole J, Lovett G, Findlay S, editors. Comparative analyses of ecosystems: Patterns, mechanisms, and theories. New York NY, USA: Springer-Verlag.

Jafvert CT, Westall JC, Grieder E, Schwarzenbach RP. 1990. Distribution of hydrophobic ionogenic organic compounds between octanol and water: Organic acids. *Environ Sci Technol* 24:1795-1803.

Jarvinen AW, Ankley GT. 1999. Linkage of effects to tissue residues: Development of a comprehensive database for aquatic organisms exposed to inorganic and organic chemicals. Pensacola FL, USA: SETAC. 364 p.

Jimenez BD, Cirmo CP, McCarthy JF. 1987. Effects of feeding and temperature on uptake, elimination, and metabolism of benzo(a)pyrene in the bluegill sunfish (*Lepomis macrochirus*). *Aquat Toxicol* 10:41-57.

Jobling M. 1981. The influence of feeding on the metabolic rate of fishes: A short review. *J Fish Biol* 18:385-400.

Johnson IC, Keller AE, Zam SG. 1993. A method for conducting acute toxicity tests with early life stages of freshwater mussels. In: Environmental toxicology and risk assessment. Landis WG, Hughes JS, Lewis MA, editors. Philadelphia PA, USA: ASTM. STP 1179. p 381-396.

Kaiser KLE, Valdmanis I. 1982. Apparent octanol/water partition coefficients of pentachlorophenol as a function of pH. *Can J Chem* 60:2104-2106.

Karr JR. 1991. Biological integrity: A long-neglected aspect of water resource management. *Ecol Appl* 1:66-84.

Keating KI. 1985. The influence of vitamin B_{12} deficiency on the reproduction of *Daphnia pulex* Leydig (Cladocera). *J Crustac Biol* 5:130-136.

Keating KI, Dagbuson BC. 1984. Effect of selenium deficiency on cuticle integrity in the Cladocera (Crustacea). *Proc Natl Acad Sci* 81:3433-3437.

Keller AE. 1993. Acute toxicity of several pesticides, organic compounds, and a wastewater effluent to the freshwater mussel *Anodonta imbicillis*. *Bull Environ Contam Toxicol* 51:696-702.

Keller AE, Ruessler DS. 1997. The toxicity of malathion to unionid mussels: Relationship to expected environmental concentrations. *Environ Toxicol Chem* 16:1028-1033.

Kiffney PM, Clements WH. 1996. Effects of heavy metals on stream macroinvertebrate assemblages from different elevations. *Ecol Appl* 6:472-481.

Kishino T, Kobayashi K. 1995. Relation between toxicity and accumulation of chlorophenols at various pH, and their absorption mechanism in fish. *Water Res* 29:431-442.

Kumaraguru AK, Beamish FWH. 1986. Effect of permethrin on the bioenergetics of rainbow trout, *Salmo gairdneri*. *Aquat Toxicol* 9:47-58.

Landrum PF, Nihart SR, Eadle BJ, Gardner WS. 1984. Reverse phase separation method for determining pollutant binding to Aldrich humic acid and dissolved organic carbon of natural waters. *Environ Sci Technol* 18:187-192.

Lanno RP, Dixon DG. 1996. The chronic toxicity of thiocyanate to rainbow trout (*Oncorhynchus mykiss*). *Can J Fish Aquat Sci* 53:2137-2146.

Lanno RP, Hickie BE, Dixon DG. 1989. Feeding and nutritional considerations in aquatic toxicology. *Hydrobiologia* 188/189:525-531.

Lanno RP, Hicks B, Hilton JW. 1987. Histological observations on intrahepatocytic copper-containing granules in rainbow trout reared on diets containing elevated levels of copper. *Aquat Toxicol* 10:251-263.

Lanno RP, Hilton JW, Slinger SJ. 1985. Maximum tolerable and toxicity levels of dietary copper in rainbow trout (*Salmo gairdneri* Richardson). *Aquaculture* 49:257-268.

Lett PG, Farmer GJ, Beamish FWH. 1976. Effect of copper on some aspects of the bioenergetics of rainbow trout (*Salmo gairdneri*). *J Fish Res Board Can* 33:1335-1342.

Lewis MA. 1995. The use of freshwater plants in phytotoxicity testing. *Environ Pollut* 87:319-336.

Liber K, Kaushi NK, Solomon KR, Carey JH. 1992. Experimental designs for aquatic mesocosm studies: A comparison of the "ANOVA" and "regression" design for assessing the impact of tetrachlorophenol on zooplankton populations in limnocorrals. *Environ Toxicol Chem* 11:61-77.

Mancini JL. 1983. A method for calculating effects, on aquatic organisms, of time varying concentrations. *Water Res* 17:1355-1362.

Markich SJ, Camilleri C. 1997. Investigation of metal toxicity to tropical biota: Recommendations for revision of the Australian water quality guidelines. Canberra, Australia: Report 127. Supervising Scientist.

Markich SJ, Jeffree RA. 1994. Absorption of divalent trace metals as analogues of calcium by Australian freshwater bivalves—An explanation of how water hardness reduces metal toxicity. *Aquat Toxicol* 29:257-290.

Mayer FL, Mehrle PM, Sanders HO. 1977. Residue dynamics and biological effects of polychlorinated biphenyls in aquatic organisms. *Arch Environ Contam Toxicol* 5:501-511.

McCarty LS, Mackay D. 1993. Enhancing ecotoxicological modeling and assessment. *Environ Sci Technol* 27:1719-1728.

Meador JP. 1991. The interaction of pH, dissolved organic carbon, and total copper in the determination of ionic copper and toxicity. *Aquat Toxicol* 19:13-32.

Mekenyan OG, Veith GD, Bradbury SP, Zaharieva N. 1995. SAR models for metabolic activation: Stability of organic cation intermediates. *Quant Struct-Act Relat* 14:264-269.

Menzer RE, Lewis MA, Fairbrother A. 1994. Methods in environmental toxicology. In: Hayes, AW, editor. Principles and methods of toxicology. 3rd ed. New York NY, USA: Raven Press. p 1391-1418.

Meyer JS, Gulley DD, Goodrich MS, Szmania DC, Brooks AS. 1995. Modeling toxicity due to intermittent exposure of rainbow trout and common shiners to monochloramine. *Environ Toxicol Chem* 14:165-175.

Miller PA, Lanno RP, McMaster ME, Dixon DG. 1993. Relative contributions of dietary and waterborne copper to tissue copper burdens and waterborne-copper tolerance in rainbow trout (*Oncorhynchus mykiss*). *Can J Fish Aquat Sci* 50:1683-1689.

Moore DRJ, Caux PY. 1997. Estimating low toxic effects. *Environ Toxicol Chem* 16:794-801.

Morel FMM. 1983. Principles of aquatic chemistry. Toronto ON, Canada: John Wiley. 446 p.

Mount DR, Swanson MJ, Breck JE, Farag AM, Bergman HL. 1990. Responses of brook trout (*Salvelinus fontinalis*) fry to fluctuating acid, aluminum, and low calcium exposure. *Can J Fish Aquat Sci* 47:1623-1630.

Munawar M, Munawar IF, Leppard GG. 1989. Early warning assays: An overview of toxicity testing with phytoplankton in the North American Great Lakes. *Hydrobiologia* 188-189:237-246.

Munns Jr WR, Gleason TR, Clancy N, Keller A, Poucher S, Lussier S. 1995. Development of population models for the risk based approach to criteria derivation: Interim report. Narragansett RI, USA: USEPA, Environmental Research Laboratory, and Aquatic Life Criteria Guidelines Committee. NHEERL-NAR Contribution nr 1726.

Neely WB. 1984. An analysis of aquatic toxicity data: Water solubility and acute LC50 fish data. *Chemosphere* 7:813-819.

Nimrod AC, Benson WH. 1998. Assessment of estrogenic activity in fish. In: Roland R, editor. Chemically induced alterations in the functional development and reproduction of fishes. Pensacola FL, USA: SETAC. p 87-100.

Norberg-King TJ, Durhan EJ, Ankley GT, Robert ED. 1991. Application of toxicity identification evaluation procedures to the ambient waters of the Colusa Basin Drain, California. *Environ Toxicol Chem* 10:891-900.

Odum EP. 1984. The mesocosm. *BioScience* 34:558-562.

[OECD] Organization for Economic Cooperation and Development. 1995. Guidance document for aquatic effects assessment. Paris, France: OECD. 112 p.

O'Hara TM, Rice CD. 1996. Polychlorinated biphenyls. In: Fairbrother A, Locke LN, Hoff GL, editors. Noninfectious diseases of wildlife. 2nd ed. Ames IA, USA: Iowa State University Press. p 71-86.

Ohlendorf HM. 1996. Selenium. In: Fairbrother A, Locke LN, Hoff GL, editors. Noninfectious diseases of wildlife. 2nd ed. Ames IA, USA: Iowa State University Press. p 128-140.

Ouellet M, Bonin J, Rodrique J, DesGranges J, Lair L. 1997. Hindlimb deformities (ectromelia, ectrodactyly) in free-living anurans from agricultural habitats. *J Wildl Dis* 33:95-104.

Pagenkopf GK. 1983. Gill surface interaction model for trace-metal toxicity to fishes: Role of complexation, pH, and water hardness. *Environ Sci Technol* 17:654-660.

Palmer TM. 1995. The influence of spatial heterogeneity on the behavior and growth of two herbivorous stream insects. *Oecologia* 104:476-486.

Pascoe D, Williams KA, Green DWJ. 1989. Chronic toxicity of cadmium to *Chironomus riparius* (Meigen)-effects upon larval development and adult emergence. *Hydrobiologia* 175:109-115.

Payne AG, Hall RH. 1979. A method for measuring algal toxicity and its application to the safety assessment of new chemicals. In: Marking LL, Kimerle RA, editors. Aquatic toxicology. Philadelphia PA, USA: ASTM. STP 667. p 171-180.

Playle RC. 1998. Modelling metal interactions at fish gills. *Sci Total Environ* 219:147-163.

Playle RC, Dixon DG, Burnison K. 1993a. Copper and cadmium binding to fish gills: Estimates of metal-gill stability constants and modelling of metal accumulation. *Can J Fish Aquat Sci* 50:2678-2687.

Playle RC, Dixon DG, Burnison K. 1993b. Copper and cadmium binding to fish gills: Modification by dissolved organic carbon and synthetic ligands. *Can J Fish Aquat Sci* 50:2667-2677.

Poff NL, Ward JV. 1990. Physical habitat template of lotic systems: Recovery in the context of historical pattern of spatiotemporal heterogeneity. *Environ Manag* 14:629-645.

Prothro MG. 1993. Office of water policy and technical guidance on interpretation and implementation of aquatic metals criteria. Washington DC, USA: USEPA, Office of Water. Memorandum from Acting Assistant Administrator for Water. 7 p; attachments 41 p.

Quinn JM, Davies-Colley RJ, Hickey CW, Vickers ML, Ryan PA. 1992. Effects of clay discharges on streams, 2. Benthic invertebrates. *Hydrobiologia* 248:235-247.

Quinn JM, Hickey CW. 1993. Effects of sewage stabilization lagoon effluent on stream invertebrates. *J Aquat Ecosys Health* 2:205-219.

Radetski CM, Ferard JF, Blaise C. 1995. A semi-static microplate-based phytotoxicity test. *Environ Toxicol Chem* 14:299-302.

Rapport DJ, Regier HA, Hutchinson TC. 1985. Ecosystem behavior under stress. *Am Nat* 125:617-640.

Resh VH, Norris RH, Barbour MT. 1995. Design and implementation of rapid assessment approaches for water resource monitoring using benthic macroinvertebrates. *Aust J Ecol* 20:108-121.

Ringer RK, Hornshaw TC, Aulerich RJ. 1991. Mammalian wildlife (mink and ferret) toxicity test protocols (LC50, reproduction, and secondary toxicity). Washington DC, USA: USEPA. EPA-600-3-91-043.

Saarikoski J, Lindstrom M, Tyynila M, Viluksela M. 1986. Factors affecting the absorption of phenolics and carboxylic acids in the guppy (*Poecilia reticulata*). *Ecotoxicol Environ Saf* 11:158-173.

Schindler DW. 1987. Detecting ecosystem responses in anthropogenic stress. *Can J Fish Aquat Sci* (Suppl):6-25.

Schindler DW, Mills KH, Malley DF, Findlay DL, Shearer JA, Davies IJ, Turner MA, Linsey GA, Cruikshank DR. 1985. Long term ecosystem stress: The effects of years of experimental acidification on a small lake. *Science* 228:1395-1401.

Schüürmann G, Schindler M. 1993. Fish toxicity and dealkylation of aromatic phosphorothionates—QSAR analysis using NMR chemical shifts calculated by the IGLO method. *J Environ Sci Health Part A Environ Sci Eng* 28:899-921.

Sibley PK, Benoit DA, Ankley GT. 1997. The significance of growth in *Chironomus tentans* toxicity tests: Relationship to reproduction and demographic endpoints. *Environ Toxicol Chem* 16:336-345.

Sprague JB. 1973. The ABCs of pollutant bioassay using fish. In: Cairns Jr J, Dickson KL, editors. Biological methods for the assessment of water quality. Philadelphia PA, USA: ASTM. STP 528. p 6-30.

Stehlik LL, Merriner JV. 1983. Effects of accumulated dietary kepone on spot (*Leiostomus xanthurus*). *Aquat Toxicol* 3:345-358.

Stephan CE, Mount DI, Hansen DJ, Gentile JH, Chapman GA, Brungs WA. 1985. Guidelines for deriving numerical national water quality criteria for the protection of aquatic organisms and their uses. Duluth MN, USA: USEPA, Environmental Research Laboratory. NTIS nr PB85 227049. 98 p.

Stephan CE, Rogers JW. 1985. Advantages of using regression analysis to calculate results of chronic toxicity tests. In: Bahner RC, Hansen DJ, editors. Aquatic toxicology and hazard assessment: 8th symposium. Philadelphia PA, USA: ASTM. STP 891. p 328-338.

Stewart-Oaten A, Murdoch WW, Parker KR. 1986. Environmental impact assessment: "Pseudoreplication" in time? *Ecology* 67:929-940.

Suter GW, Rosen AE, Linder E, Parkhurst DF. 1987. Endpoints for responses of fish to chronic toxic exposures. *Environ Toxicol Chem* 6:793-809.

Swartz RC, Ferraro SP, Lamberson JO, Cole FA, Ozretich RJ, Boese BL, Schults DW, Behrenfeld M, Ankley GT. 1997. Photoactivation and toxicity of mixtures of polycyclic aromatic hydrocarbon compounds in marine sediment. *Environ Toxicol Chem* 16:2151-2157.

Swartz RC, Schults DW, Ozretich RJ, Lamerson JO, Cole FA, DeWitt TH, Redmond MS, Ferraro SP. 1995. PAH: A model to predict the toxicity of polynuclear aromatic hydrocarbon mixtures in field collected sediments. *Environ Toxicol Chem* 14:1977-1987.

Thursby G, Miller DC, Poucher S, Coiro L, Munns W, Gleason T. 2000. Ambient water quality criteria for dissolved oxygen (saltwater): Cape Cod to Cape Hatteras. Washington DC, USA: USEPA, Office of Water. EPA-822-D-99-002.

[USEPA] U.S. Environmental Protection Agency. 1984. Ambient water quality criteria for copper. Washington DC, USA: USEPA, Office of Water, Regulations and Standards. EPA-440-5-84-031. NTIS PB85-227023.

[USEPA] U.S. Environmental Protection Agency. 1989. Short-term methods for estimating the chronic toxicity of effluents and receiving waters to freshwater organisms. 2nd ed. Cincinnati OH, USA: USEPA, Environmental Monitoring Systems Laboratory. EPA-600-4-889-001.

[USEPA] U.S. Environmental Protection Agency. 1994. Interim guidance on determination and use of water-effect ratios for metals. EPA-823-B-94-001.

[USEPA] U.S. Environmental Protection Agency. 1995a. Environmental monitoring and assessment program for surface waters: Field operations and methods for measuring the ecological condition of wadable streams. Lazorchak JM, Klemm DJ, editors. Washington DC, USA: USEPA. EPA-620-R-95-004.

[USEPA] U.S. Environmental Protection Agency. 1995b. Great Lakes water quality initiative: Technical support document for the procedure to determine bioaccumulation factors. Washington DC, USA: USEPA. EPA-820-B-95-005.

[USEPA] U.S. Environmental Protection Agency. 1998. Guidelines for ecological risk assessment. Risk assessment forum. Washington DC, USA: USEPA. EPA-630-R-95-002F. p 28.

Van den Berg CMG, Wong PTS, Chau YK. 1979. Measurement of complexing materials excreted from algae and their ability to ameliorate copper toxicity. *J Fish Res Board Can* 36:901-905.

Van Leeuwen CJ, Hermens JLM. 1995. Risk assessment of chemicals: An introduction. Dordrecht, The Netherlands: Kluwer.

Van Tilborg WJM, van Assche F. 1998. Homeostatic regulation defines a stress-free concentration band for essential elements relevant for risk assessment. *SETAC News* 18(3):27-28.

Van Veld PA, Stegeman JJ, Woodin BR, Patton JS, Lee RF. 1988. Induction of monooxygenase activity in the intestine of spot (*Leiostomus xanthurus*), a marine teleost, by dietary PAH. *Drug Metab Dispos* 16:659-669.

Van Wezel AP, Opperhuizen A.1995. Narcosis due to environmental pollutants in aquatic organisms: Residue-based toxicity, mechanisms, and membrane burdens. *Crit Rev Toxicol* 25:255-279.

Veith GD, Kosian PA. 1982. Estimating bioconcentration potential from octanol/water partition coefficients. In: MacKay D, Paterson S, Eisenreich SJ, Simmons MS, editors. Physical behavior of PCBs in the Great Lakes. Ann Arbor MI, USA: Ann Arbor Science.

Walker MK, Cook PM, Butterworth BC, Zabel EW, Peterson RE. 1996. Potency of a mixture of polychlorinated dibenzo p dioxin, dibenzofuran, and biphenyl congeners compared to 2,3,7,8 tetrachlorodibenzo p dioxin in causing fish early life stage mortality. *Fundam Appl Toxicol* 30:178-186.

Wallace JB, Vogel DS, Cuffney TF. 1986. Recovery of a headwater stream from an insecticide-induced community disturbance *J North Am Benthol Soc* 5:115-126.

Walsh GE, Duke KM, Foster RB. 1982. Algae and crustaceans as indicators of bioactivity of industrial wastes. *Water Res* 16:879-883.

Watt PW, Marshall PA, Heap SP, Loughna PT, Goldspink G. 1988. Protein synthesis in tissues of fed and starved carp, acclimated to different temperatures. *Fish Physiol Biochem* 4:165-173.

Welsh PG, Skidmore, JF, Spry DJ, Dixon DG, Hodson PV, Hutchinson NJ, Hickie BE. 1993. Effect of pH and dissolved organic carbon on the toxicity of copper to larval fathead minnow (*Pimephales promelas*) in natural lake waters of low alkalinity. *Can J Fish Aquat Sci* 50:1356-1362.

West CW, Ankley GT, Nichols JW, Elonen GE, Nessa DE. 1997. Toxicity and bioaccumulation of 2,3,7,8-tetrachlorodibenzo-p-dioxin in long-term tests with the freshwater invertebrates *Chironomus tentans* and *Lumbriculus variegatus*. *Environ Toxicol Chem* 16:1287-1294.

White A, Fletcher TC. 1986. Serum cortisol, glucose and lipids in plaice (*Pleuronectes platessa* L.) exposed to starvation and aquarium stress. *Comp Biochem Physiol* 84A:649-653.

Wiens JA, Parker KR. 1995. Analyzing the effects of accidental environmental impacts: Approaches and assumptions. *Ecol Appl* 5:1069-1083.

Wigglesworth VB. 1970. Insect hormones. San Francisco CA, USA: WH Freeman.

Winner RW, Boesel MW, Farrell MP. 1980. Insect community structure as an index of heavy metal pollution in lotic ecosystems. *Can J Fish Aquat Sci* 37:647-655.

Wong PTS, Chau YK, Luxon PL. 1978. Toxicity of a mixture of metals on freshwater algae. *Can J Fish Aquat Sci* 35:479-481.

Wood CM. 1992. Flux measurements as indices of H^+ and metal effects on freshwater fish. *Aquat Toxicol* 22:239-264.

Wood CM, Playle RC, Hogstrand C. 1999. Physiology and modelling the mechanism of silver uptake and toxicity in fish. *Environ Toxicol Chem* 18:71-83.

Woodward DF, Brumbaugh WG, DeLonay AJ, Little EE, Smith CE. 1994. Effects on rainbow trout fry of a metals-contaminated diet of benthic invertebrates from the Clark Fork River, Montana. *Trans Am Fish Soc* 123:51-62.

Woodward DF, Farag AM, Bergman HL, DeLonay AJ, Little EE, Smith CE, Barrows FT. 1995. Metals-contaminated benthic invertebrates in the Clark Fork River, Montana: Effects on age-0 brown trout and rainbow trout. *Can J Fish Aquat Sci* 52:1994-2004.

Zabel EW, Cook PM, Peterson RE. 1995. Toxic equivalency factors of polychlorinated dibenzo-*p*-dioxin, dibenzofuran and biphenyl congeners based on early life stage mortality in rainbow trout (*Oncorhynchus mykiss*). *Aquat Toxicol* 31:315-328.

Zacharewski T. 1997. In vitro bioassays for assessing estrogenic substances. *Environ Sci Technol* 31:613-623.

CHAPTER 4

Risk Characterization

Dwayne R.J. Moore, Charles G. Delos, Jeffrey M. Giddings, David J. Hansen, Wayne G. Landis, Eugene R. Mancini, Beth L. McGee, Keith R. Solomon, John E. Toll, William Wuerthele, William J. Adams

Introduction

This workshop recommended that quantitative ERA procedures be used to effectively integrate water-quality management objectives with WQC derivation procedures to improve the overall process. To appropriately derive protective criteria, specific water-quality management objectives must be established (e.g., designated water body uses) for the development, modification, and use of such criteria. Within this framework, ERA provides a unique set of tools for quantifying risk and ensuring that a given level of protection is provided. Risk characterization is the part of the risk assessment process in which exposure and effects data are integrated in a manner that provides a statement of the probability of effects occurring at a given exposure concentration. Ecological risk characterization can be described as a 2-step process. The first step is to estimate the likelihood of adverse ecological effects through the use of exposure and effects data. The second is to document and report on the nature and extent of ecological risk in an appropriate form and context for use in decision-making. It is envisioned that the application of risk assessment principles to WQC would be used to establish 3 types of criteria (see Chapter 1, Figure 1-4). Each criterion type reduces uncertainty about the values without sacrificing the level of protection. Application of risk assessment principles to WQC would be best applied in conjunction with developing site-specific criteria, that is Type 3 criteria.

For circumstances where Type 1 criteria are incorrect because of site-specific conditions or other modifying factors that lead to a high degree of uncertainty, Type 2 criteria may be derived. Incorporation of site-specific environmental factors and effects data on local species may lead to a Type 2 criterion based upon reductions in uncertainty around the established Type 1 level of protection (e.g., WER determination or recalculation procedure). Type 2 criteria may result in a greater or smaller concentration than Type 1 criteria.

In summary, the Risk Characterization Working Group concluded the following:

- Quantitative ERA procedures can be used to effectively integrate water-quality management objectives with WQC derivation procedures to improve the overall process.

- Quantitative ERA procedures might best be used in developing Type 3 criteria. Development of Type 3 criteria usually requires a site-specific management goal that replaces the generic national goal of "protecting 95% of the species 95% of the time."

- Type 3 criteria rely upon site-specific data. The most direct test to determine whether or not a criterion is protective is to conduct appropriate field studies to evaluate the health of the local indigenous populations. In situ toxicity tests can be used to help assess whether WQC are being over- or underprotective.

- The sources of uncertainty inherent in the WQC derivation process are not likely to be additive or multiplicative, that is, some uncertainties cancel out. The current procedures used for deriving WQC at the national level are biased towards overprotection (e.g., use of conservative default values for parameters affecting toxicity, use of tests on individuals, default assumptions on averaging period for acute and chronic criteria and return frequencies). Use of site-specific criteria reduces uncertainty because we are better able to adjust for the environmental conditions affecting toxicity (i.e., reduce uncertainty due to extrapolation from the laboratory to field conditions) and resident species (i.e., reduce uncertainty due to species-to-species extrapolation).

- Good communication facilitates the transfer of information and conveys the results achieved in risk characterization. In writing and communicating WQC, particular attention should be paid to using good graphics to explain relationships in the data; minimizing technical jargon, acronyms, and complex compilations of data in tables; using explanatory text boxes to flag important issues and interpretations of the data; using simple and appropriate analogies to explain complex and potentially controversial issues; and relating the WQC to the designated water use and environmental management issues.

Unique Aspects of the Type 3 Criteria Development Process

If, for a particular water-quality management issue, there is concern about the validity or level of uncertainty about a Type 2 criterion, then there may be a need to develop a Type 3 criterion. In Type 3 criteria, site-specific evaluations of biological and physical factors, microcosm test results, and field data may be used to develop a criterion that has a low level of uncertainty while achieving the desired level of protection. One of the most significant differences between Type 3 criteria and Type 1 and 2 criteria is that a site-specific management goal is required in Type 3. The Type 3 management goal replaces the generic national goal of "protecting 95% of the species 95% of the time."

The current U.S. practice is to pick the output of the WQC derivation procedure, such as the CMC or the CCC, and to use only this information for setting a water quality standard. This approach assumes that any value greater than the selected CMC or CCC presents an unacceptable risk without describing the potential severity of the risk as the concentration increases.

Establishing a management goal is an essential part of the Type 3 criterion derivation process for 2 reasons:

1) Design—The management goal provides the basis for deciding what information will be needed to derive the WQC. Therefore, it should be the basis for sampling design, assessment and measurement endpoint selection, and exposure and effects characterization methods. The management goal also determines the types of people who should be involved in the risk characterization, both for providing technical expertise (i.e., risk assessment and peer review) and for incorporating societal values (i.e., stakeholders).

2) Interpretation—Once the data are collected, models constructed and calibrated, and the exposure and effects characterizations completed, the management goal sets the agenda for the risk characterization, that is, it specifies the questions risk managers need answered by the technical experts and stakeholders. A case study of how to develop a management goal is presented later in this chapter.

Derivation of a Type 3 criterion also depends on additional lines of evidence to supplement the information collected previously. Additional lines of evidence can include the results of receiving water and sediment toxicity tests, biological surveys, and biomarker studies. The use of independent lines of evidence is an established technique that increases the reliability of risk characterization (Geckler et al. 1976; Mount et al. 1984, 1985; Mount and Norberg-King 1985, 1986; Mount, Norberg-King, Steen 1986; Mount, Steen, Norberg-King 1986a, 1986b; Norberg-King and Mount 1986; Arthur 1988; Hansen 1989).

In developing a Type 3 criterion, water-quality managers are encouraged to use all available effects characterization data to conceptualize the risk posed by the substance. To assist in this task, the criterion document should present the effects information in a more understandable format than is currently used. For example, the acute and chronic species sensitivity distributions for plants and animals should be presented in graphical as well as tabular format. This would allow the user to quickly identify sensitive species and endpoints and to make comparisons between substances. The consequences of criteria exceedances depend on the steepness of the distribution of species effect values (Figure 4-1). In this figure, a concentration of 0.1 units would result in an exceedance of species effects values for 58% of biota for substance 1, but only 16% of biota for substance 2. Both substances have the same WQC (i.e., the lines cross at the 5% effect level). Thus, distribution slopes can provide information to water-quality managers about the potential consequences of substance exceedances. Knowledge about dose–response can lead to selection of specific taxa of organisms for further evaluation and the ability to define the level of risk for a given population that may exist at a specific site.

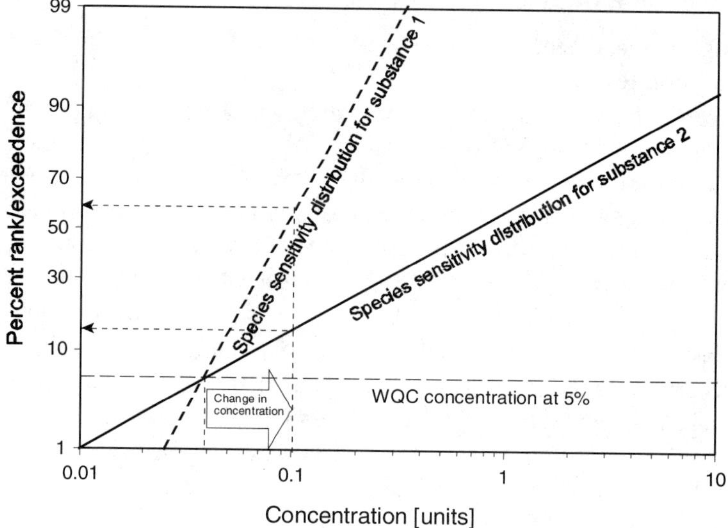

Figure 4-1 Graphical illustration of the importance of the slope of the species sensitivity distribution in relationship to changes in exposure concentration. (The concentration that affects 5% of the species [i.e., protects 95% of the species] is the same for substances 1 and 2. However, an exceedance of this concentration affects more species for substance 1 than for substance 2. The same concept is applicable to interpreting dose-response data for a given species.)

Toxicity tests with receiving waters are used to determine whether the water or sediments contain chemical concentrations toxic to the species tested. This information provides an independent site-specific measure of toxicity for comparison to generic test results conducted using standard water or sediment conditions. Because water column chemicals in most aquatic systems are transported to sediments over long periods (primarily through deposition of particles), sediment measurements can provide valuable information on the long-term conditions of a water body and the effects of cumulative chemical loading. The results of water column tests may represent transient conditions and may not be indicative of long-term quality of the aquatic environment.

Biosurveys involve sampling and counting the organisms present in a receiving environment. Benthic macroinvertebrate and fish surveys are commonly used to assess quality of the aquatic environment. Measurements of biomarkers such as biochemical, physiological, and histological changes in organisms can be used to estimate either exposure or likely effects (Huggett et al. 1992). A biomarker response that has been linked experimentally to a particular substance can provide direct evidence of exposure and, in some cases, the extent of exposure.

Multiple lines of evidence are appropriate for a Type 3 criterion because each individual line of evidence has shortcomings. For example, biosurveys are less useful than the other lines of evidence for separating the impact of a particular stressor from confounding factors, because of natural variability and the difficulty of designing controlled field experiments. Toxicity tests have laboratory-to-field and interspecies extrapolation problems. Biomarkers generally have not been linked to effects that pose acute or chronic risks to populations. Hence, multiple lines of evidence are essential to develop a Type 3 criterion that we can be confident is appropriately conservative.

Derivation of WQC inevitably involves extrapolations from laboratory studies to the field and extrapolation across species due to the limited amount of available information (Suter et al. 1985, 1993). Even for a well-studied chemical such as phenol, the species tested constitute less than 1% of the North American icthyofauna (Suter et al. 1985). Current U.S. WQC and Canadian WQGs explicitly adopt the laboratory-to-field extrapolation approach (e.g., see sections addressing applicability of national WQC to specific sites and national WQC versus field-microcosm studies in Hansen 1989). Many studies have shown that laboratory-derived WQC are protective in field situations (Geckler et al. 1976; Mount et al. 1984, 1985; Mount and Norberg-King 1985, 1986; Mount, Norberg-King, Steen 1986; Mount, Steen, Norberg-King 1986a, 1986b; Norberg-King and Mount 1986; Arthur 1988; Hansen 1989). Field testing for most chemicals is impractical due to the questionable ethics of wide-scale field testing of toxicity or locating ecosystems exposed only to the stressor of interest. Consequently, the scientific and regulatory community relies on laboratory tests to determine chemical toxicity. Laboratory tests also are used because the laboratory setting provides more control over other confounding stressors such as predation, malnutrition, competition from other species, season, habitat change, and other chemical stressors. For birds and mammals, uncertainty is reduced by limiting the taxonomic distance involved in the extrapolation (e.g., extrapolating from birds to birds, and not from birds to mammals).

Ecological Risk Assessment to Manage Water Quality: Duwamish Estuary Case Study

The King County Office of the Washington Department of Natural Resources (DNR, Washington State, USA) undertook a water-quality assessment of the Duwamish Estuary, using site-specific ERA as the methodology. The objectives of the project were to determine combined sewer overflow (CSO) impacts to the Duwamish River and Elliott Bay, Washington and to understand the possible benefits that could be achieved through CSO control. These objectives were part of the larger King County management goal of designing a CSO control strategy that continuously protects and improves water, sediment, and habitat quality in the Duwamish Estuary. DNR and Duwamish Estuary stakeholders identified 2 indicators to measure progress toward achieving the management goal: 1) abundant, diverse, and healthy biological communities; and 2) enhanced recreational, commercial, and cultural use of Duwamish River and Elliott Bay resources.

As DNR formulated the approach to be used for the water-quality assessment, it became apparent that they needed to understand not only scientific facts but also societal values. It also became apparent that the effects on biota and the identity of the causal stressors were highly uncertain, and the values subject to disagreement. Therefore, the approach had to be designed to identify and reduce major sources of uncertainty where possible and to separate matters of science from matters of value. The scientific information DNR needed to gather included descriptions of

- the biological communities and water and sediment quality of the Duwamish River and Elliott Bay,
- how water and sediment quality affect the biological communities, and
- how CSOs affect water and sediment quality.

To better understand societal values, DNR asked stakeholders what they thought about the possible benefits of CSO control, including enhanced recreational, commercial, and cultural uses of Duwamish River and Elliott Bay resources.

As this case study illustrates, selecting and defining a management goal requires input from scientists and stakeholders. Once a management goal has been defined and agreed upon, Type 3 criteria can be derived for the media and stressors of concern. Type 3 criteria can be compared to monitoring study results to determine existing water and sediment quality and can be used as performance indicators following implementation of risk management measures (e.g., CSO control).

Relationship of WQC to Community Structure and Function

There are several ways of testing the robustness of a criterion and decreasing the uncertainty in whether the level of protection is being achieved. Experiments can be conducted to test the organismal basis of the derivation, the assumptions about exposure, and the ability to protect model ecological systems and natural systems. WQC have been tested in microcosms, mesocosms, and field studies with the purpose of evaluating their relevance to field conditions. Testing the assumptions of averaging period, return frequencies, and site-specific considerations, however, have had little or no consideration. This section provides a brief overview of methods that can be used to validate the efficacy of WQC.

Single-species toxicity tests in the laboratory

Sensitivity curves and ACRs often are used to derive WQC. Additional laboratory acute and chronic toxicity tests can be used to test the robustness of the derived distributions. Species of special concern can be tested to ensure that WQC are being protective of important resources (e.g., fisheries).

Other organismal endpoints also can be examined to test the assumption that classical acute and chronic data are protective of endpoints critical to the protection of ecosystem structure and function. Potential endpoints include disruption of endocrine, immune, or behavioral activities—tests are available for some, but not all, species. Other endpoints important to species survivorship, such as nutrient uptake, migration rates, colonization ability and nesting success, can be used to evaluate the potential for toxicity and are best measured in field studies.

Multispecies toxicity tests

A variety of multispecies toxicity tests have been designed to model specific environments or specialized attributes of ecological structures. Pond-type microcosms or mesocosms as originally developed for pesticide testing (e.g., FIFRA) are designed to have some of the attributes of freshwater ponds and lakes. A variety of stream designs have been implemented to model systems with high water velocities. Less frequently used are estuarine or marine systems, perhaps due to the difficulties of successfully scaling down many of the important features of saltwater environments such as tides and currents. The Marine Environmental Research Laboratory facility at Narragansett, Rhode Island, USA, is an example of a marine model ecosystem, although replication is limited due to the size of the individual mesocosm structures.

Important endpoints that may be examined in mesocosm studies includes
- chemical exposure and fate,
- populations—numbers and dynamics,
- community structure—composition and distribution of species,
- function—nutrient fixation and flow,
- production,
- scaling features,
- heterogeneity in space and time,
- colonization sources and migration rates, and
- landscape type features.

Well-designed multispecies toxicity tests are useful tools to assist in determining the degree of protection afforded by Type 1, 2, and 3 criteria.

Field monitoring and evaluation

The most direct test of a criterion is to conduct appropriate field studies. These studies may be incident reports of alterations in receiving waters or direct investigation of receiving systems. Reports of changes in species distributions or abundances, incidents of fish kills, or invasion of exotics due to the elimination of indigenous species may all be considered as evidence that environmental quality has been degraded. If the spatial and temporal patterns of change are correlated with concentrations of a chemical stressor and the concentrations at which effects begin to appear are below the WQC for that chemical, there is indirect evidence that the criterion may not be adequately protective.

In situ toxicity tests can be used to test whether WQC are being over- or underprotective. For example, if the concentrations of all stressors of concern are below their respective WQC, but in situ tests indicate severe effects, there is evidence that one or several criteria are underprotective, or that a material with no set criterion is responsible. Conversely, if levels of some stressors of concern are well above their respective WQC, but in situ tests reveal no toxicity, there is evidence that some of the criteria are overprotective. In both cases, the evidence is circumstantial only, and not useful for determining which criteria require adjustment or by how much.

Although confounding stressors exist, the ultimate verification of the performance of the WQC is an in-depth investigation of the structure and function of the receiving waters. Advances in sampling schemes, data analysis tools, and new paradigms describing ecological relationships have dramatically improved our ability to detect alterations and attribute causality (e.g., Landis et al. 1994, 1996, 1998).

One strategy to overcome the problem of detecting a contaminant-related pattern among those caused by other stressors is to use a gradient approach. Such an approach does not rely on an upstream or downstream comparison or require the designation of a reference site. Instead, samples are taken in a variety of areas matched to contaminant concentration or other physical characteristics, and the data are searched for patterns. The samples should be taken several times a year so that gradients that occur seasonally can be detected and accounted for in the analysis. Such an analysis of benthic invertebrate populations was performed by Matthews et al. (1991) for 2 streams in Washington State, with one having input from a wastewater treatment plant. Multivariate analysis was used to search for and analyze patterns within the data set. One of the patterns, using principal components analysis (PCA), was dependent upon the time of year (season) in which the data were taken. PCA attempts to explain variance, and seasonal changes certainly are a major contributor to sample variance in both of the study streams. Another technique, nonmetric clustering, detected a pattern related to distance from the treatment outflow. Nonmetric clustering attempts to categorize data sets without regard to distribution or relative variance. In this study, the seasonal pattern was separated from the pattern associated with the treatment plant outfall.

Considerations of Ecological Significance in Risk Characterization

The purpose of WQC is often understood to be protection of "most of the species, most of the time." However, the fraction of species affected is a simplistic measure of "ecological health" or "environmental integrity." If critical species in an ecosystem are affected by a chemical, the function of the ecosystem may be impaired to the extent that other species are affected indirectly and the overall ecological impact amplified. On the other hand, an ecosystem can tolerate effects on some species if other species that perform the same ecological function are unaffected (Figure 4-2). Thus, a given protection goal, such as 95% of the species, may be over- or underprotective of the desired ecological state of a particular water body, depending on the ecological role of the most sensitive species. (Threatened and endangered species often require special consideration; see discussion in "Threatened and endangered species and organisms of special concern," p 137.) Higher criteria types should be written to address ecological and management goals in addition to the goal of protecting species. This section discusses some of the ecological factors that should be taken into account.

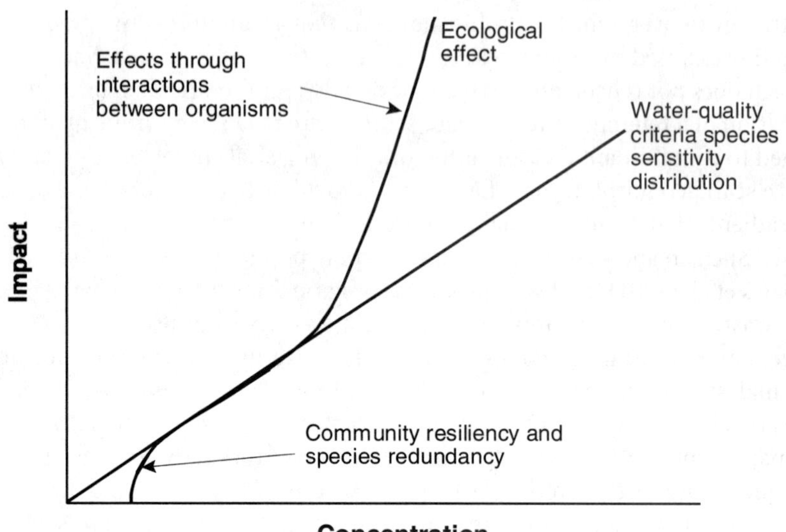

Figure 4-2 The conceptual relationship between chemical concentration, number of species affected, and ecological impact. At high exposures, the number of species directly affected by a chemical is so large that the system can no longer support the more tolerant species. At low exposures, effects on sensitive species are compensated for by community resiliency and species redundancy. The exposure concentrations at which effects are amplified or compensated vary, depending on the chemical and the ecosystem.

Ecological redundancy

From an ecological point of view, effects on a population are not necessarily of concern as long as the functions the population performs can be taken over by other species. The term "function" here is used loosely to refer to any of the population's interactions with other populations or with the abiotic environment. The functions of interest are mainly related to energy and nutrient flow: producing new biomass (primary production), harvesting biomass; controlling the abundance of lower trophic levels, providing food to higher trophic levels, or processing organic detritus (e.g., shredding plant tissue, macerating animal remains, and mineralizing organic compounds).

Aquatic ecosystems, especially in the temperate zone, exhibit "functional redundancy" (Walker 1992, 1995; Baskin 1994), that is, multiple species are present to perform each critical function. In most aquatic ecosystems, for example, there are 3 major plant communities (phytoplankton, periphyton, and macrophytes), each of which may include dozens of species. Each plant community supports a diverse group of herbivores. Many microinvertebrates and macroinvertebrates are generalized or specialized detritivores. Invertebrate and vertebrate predators in aquatic systems usually are opportunistic and will feed on a wide variety of prey, depending on what is available. This functional redundancy is critical to the persistence of the ecosystem as a whole and is the product of evolutionary adaptation to fluctuating environmental conditions.

Functional redundancy is highly relevant to the evaluation of ecological effects of toxic substances because it implies that reductions in a few sensitive populations are unlikely to impair the functions of the ecosystem. Although one can hypothesize a chain of events whereby toxic effects on a population cause ripples throughout the ecosystem, observations in mesocosms and in field studies tell us the opposite is more common—a community or ecosystem is less sensitive than its most sensitive populations (e.g., Moore 1998), just as a population is less sensitive than its most sensitive individuals. Thus, for example, the loss of amphipods and isopods from mesocosms exposed to cypermethrin had no apparent impact on the ecosystem overall; several groups of organisms, including rotifers, chydorid copepods, chironomids, oligochaetes, and snails, expanded to fill the functional role of the missing populations (Farmer et al. 1995). Where critical populations are affected, however, the opposite may be true. For example, removal of *Mysis relicta*, a voracious zooplanktivore, from lakes causes dramatic changes in zooplankton populations which, in turn, cause major changes in phytoplankton communities via trophic cascading (Schindler 1996).

Controlled exposure–response experiments with microcosms and mesocosms have demonstrated that, at some level of exposure, temporary changes occur in the population abundance of a few, highly sensitive species. At a higher level of exposure, more severe and longer-lasting effects occur to a larger number of taxa. The transition from minor (ecologically tolerable) impact to ecologically significant impact usually occurs at exposure concentrations greater than the 10th percentile of single-species acute or chronic toxicity values (Figure 4-3; Giddings et al. 1996, 1997; Solomon et al. 1996; Versteeg et al. 1999).

Ecological role of sensitive species

To understand the ecological significance of chemical effects, the ecological role of the sensitive species must be taken into account. For example, cladoceran crustaceans are extremely sensitive to organophosphate insecticides. An assessment of the impact of the organophosphate diazinon in the San Joaquin River, California, USA concluded that, although diazinon concentrations are high enough at certain times and places to cause mortality of cladocerans, this group of organisms is not of critical importance to fish or other components of the San Joaquin ecosystem (Giddings et al. 1997). In the ecological context, the cladocerans were not keystone species. In contrast, the decreasing abundance of alligators in the southeastern U.S. has led to serious habitat disruption because the activities of alligators create water holes upon which many other species depend. In this situation, the alligators are keystone species.

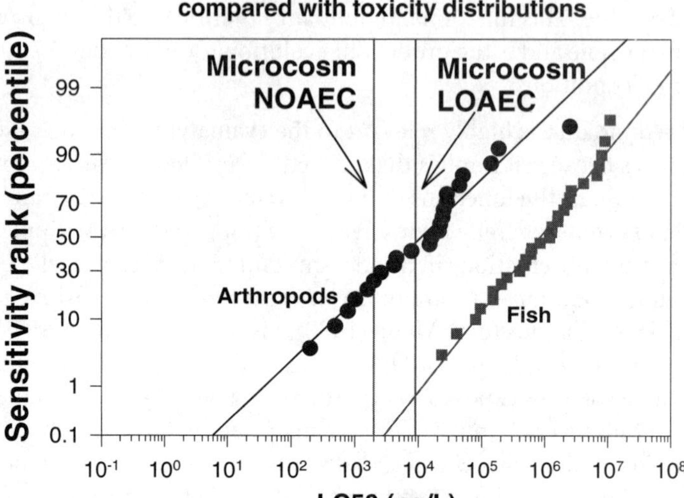

Figure 4-3 Example of microcosm data indicating a threshold of
ecologically significant effects (NOEC and lowest-observed-adverse-effect
concentration [LOAEC]) above the 10th percentile of single species LC50)
concentration for arthropods (the most sensitive taxonomic group)
(Giddings et al. 1997)

Extrapolating effects from individuals to populations

Water-quality criteria are derived from measurement endpoints, such as LC50 and
ECx values that reflect the magnitude of effects on individuals. Measurement of
percent mortality in a laboratory toxicity test does not give direct information
about the magnitude of effects on natural populations. A variety of population
models are available that can be used, especially in derivation of Type 3 criteria, to
extrapolate from percent mortality to population endpoints such as percent
reduction in abundance, time to recovery, or likelihood of local population
extinction (Barnthouse et al. 1988; Suter 1993; Caswell 1996). Simple techniques
such as life-table analysis can be used to derive generic relationships for representative life history strategies to convert percent mortality (or other toxicity test
endpoint) to population endpoints. More sophisticated models such as those used
for fisheries management, population distribution models, and metapopulation
models, which require a great deal of data for accurate predictions, can be used to
estimate effects on specific populations at specific sites.

Ecological recovery

Effects of chemicals on a population or ecosystem are of less ecological significance if recovery can take place within a reasonably short period of time. "Recovery" can be defined as return of an ecological parameter (such as the abundance of

a population or the number of species in a community) to the normal range of an undisturbed system. Knowledge of the ecology and life history of affected species is necessary for evaluation of recovery potential. Population models can be useful in estimating the time required for recovery.

Several factors contribute to the recovery of populations after an impact such as exposure to a toxic substance (Wallace 1990; Yount and Niemi 1990; Detenbeck et al. 1992; Whiles and Wallace 1992, 1995; Tikkanen et al. 1994). Most aquatic invertebrates have rapid growth rates and short generation times and are capable of rapid population growth. Resistant life stages allow many aquatic invertebrate species to become reestablished when toxicant concentrations decline after periods of high exposure. Spatial variability of concentrations in water or sediment creates low-exposure refuges from which recolonization can occur. Immigration from nearby unexposed areas also can lead to rapid reestablishment of affected populations (see "Landscape considerations," below). Insect recolonization occurs through egg deposition by adults from outside the affected ecosystem.

The time for recovery is clearly important in establishing the acceptable frequency of WQC exceedances. If the sensitive species upon which the criterion concentration is based have a high recovery potential (such as phytoplankton, zooplankton, and many macroinvertebrates), exceedances can be more frequent than if the sensitive species are long-lived and slow to recover (such as predatory fish).

Landscape considerations

The degree of change and the persistence of the alteration in a system are tied to a variety of landscape features specific to the environment under consideration:

- Regions of small area will have fewer species than larger ones. Rates of species turnover also will be higher.
- Areas nearer sources of colonization generally have a faster rate of recolonization and population recovery than more isolated areas.
- Recolonization has a large stochastic component. Although the number of species may be reestablished, the types may be different due to chance events and may not be characteristic of the contaminant.
- Impacts to parts of the landscape that are colonization sources (main river channels, lakes, estuaries, breeding habitats, etc.) will be more wide-ranging and persistent than impacts to colonization sinks.
- Contamination of a site that is critical to the migration of organisms in a landscape can have disproportionate impacts to the numbers and distributions of species in the landscape.
- Habitat and landscape considerations may outweigh contaminant characteristics as the final determinant of the magnitude and duration of the impact.

Characterizing Uncertainty

A key part of the risk characterization phase for WQC is to provide a summary of the assumptions, scientific uncertainties, and strengths and limitations of the analyses. There are many ways to characterize uncertainty. They range from a simple listing of sources of uncertainty to quantitative methods such as first-order error propagation, Monte Carlo simulation, probability bounds analysis, and others. This section will briefly describe the major sources of uncertainty for WQC, and then provide guidance on how to characterize uncertainty for Type 1, 2, and 3 criteria.

Sources of uncertainty

There are a number of sources of uncertainty inherent in the derivation of WQC. Several of these uncertainties arise from the extrapolations made during the derivation of criteria, including the following:

- Species-to-species extrapolation. WQC are based on the results of toxicity tests carried out on standardized bioassay species (e.g., rainbow trout, daphnids, *Selenastrum*). In Canada, the most sensitive species in the toxicity data set is generally used as the basis for deriving a WQG. In the U.S., the goal is protection of 95% of the species 95% of the time. To consistently achieve this, the concentration that protects 95% of the tested species is mathematically derived. In either case, there is the implicit assumption that the sensitivities of the chosen bioassay species are representative of the sensitivities of species in the field. For jurisdictions that use the most sensitive species as the basis for deriving a criterion, it is highly unlikely that the most sensitive species in nature will be among those routinely tested in the laboratory. For those jurisdictions that rely on a species sensitivity distribution, there is the distinct possibility that the species routinely used in toxicity tests (particularly rainbow trout and *Daphnia*) were chosen as standard bioassay species because they survived well in the laboratory and were convenient to test. This would mean that species sensitivity distributions might not be representative of species sensitivities in nature.

- Laboratory versus field conditions. Most laboratory bioassays are designed to maximize bioavailability of the test substance. For example, concentrations of humic acids in test waters are very low, thus increasing the bioavailability of metals and nonpolar organic chemicals over what would be the case in most field situations. Laboratory concentrations are usually constant, whereas chemical exposures in the field may or may not be constant. Further, biota in nature often are exposed to other stressors (e.g., other chemicals, disease, competition from other species) that increase their susceptibility to the chemical of interest. The net result of these and other differences between laboratory and field conditions is that uncertainties exist about the appropriateness of the extrapolation.

- Lower to higher levels of biological organization. The level of protection provided by WQC is usually stated as protection of the most sensitive species or most of the species most of the time. Therefore, protection is aimed at the population level or perhaps the community level of organization. Most toxicity tests used in criteria derivation, however, are carried out on individuals of a particular species. To translate effects on individuals to populations requires consideration of temporal and spatial patterns of exposure and life history parameters such as mortality and fecundity for different age classes of the population of interest. To translate effects to the community level requires knowledge about species interactions with each other and with the abiotic environment. Such analyses are rarely, if ever, done in the derivation of WQC. Hierarchy theory suggests that this source of uncertainty is biased towards overprotection because effects can occur at lower levels of organization without being transmitted to higher levels, while the reverse situation cannot occur (Allen and Starr 1982; O'Neill et al. 1986; Moore 1998).

There are many other sources of uncertainty in addition to the extrapolation sources discussed above. Rarely are we able to account for uncertainties arising from, for example, human error (e.g., calculation and transcription errors), intratest and intertest variability, and lack of knowledge about how ecosystems function.

The sources of uncertainty inherent in the WQC derivation process are not likely to be additive or multiplicative. That is, some uncertainties cancel out. The current procedures used for deriving WQC at the national level are biased towards over-protection (e.g., use of conservative default values for parameters affecting toxicity, use of tests on individuals, default assumptions on averaging period for acute and chronic criteria, and return frequencies). Use of site-specific criteria reduces uncertainty because we are better able to adjust for the environmental conditions affecting toxicity (i.e., reduce uncertainty due to extrapolation from the laboratory-to-field conditions) and the resident species (i.e., reduce uncertainty due to species-to-species extrapolation). Further, use of other lines of evidence (e.g., mesocosm and field trial results, field observations, effects of related chemicals in the field) can reduce the uncertainty about whether a WQC is achieving the desired level of protection.

The remainder of this section briefly describes how uncertainty should be characterized for Type 1, 2, and 3 criteria, and the benefits of doing so.

Type 1 and 2 criteria

Uncertainty about the degree of over- or underprotection of a WQC will be greatest for Type 1 criteria (Figure 4-4). Because of the conservatism built into the derivation process for national WQC, the actual level of protection achieved will be unknown but may substantially exceed the stated desired level of protection (e.g., protection of 95% of biota). Uncertainty about the level of protection achieved with Type 1 WQC at a site, however, would be difficult to quantify without a probabilistic risk assessment. Nevertheless, there are several procedures that can be used to acknowledge, quantify, and reduce the uncertainties that exist.

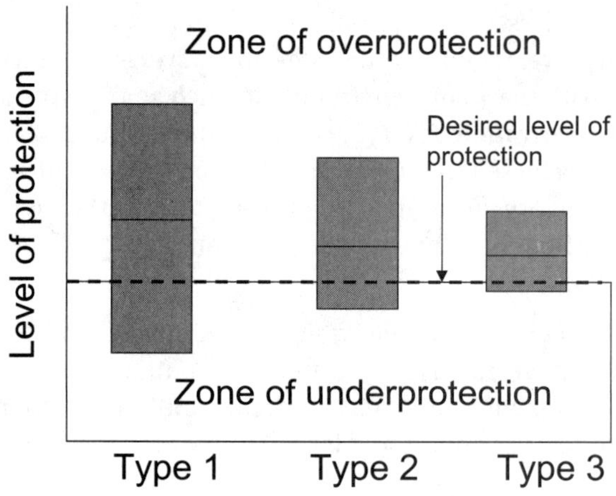

Figure 4-4 Relationship between level of protection provided by each criteria type and uncertainty associated with the level of protection. The desired level of protection remains constant in each criteria type, but the uncertainty is reduced in Types 2 and 3 as compared to Type 1. Mean of uncertainty ranges depicted as line in each box.

Defining WQC as precise point estimates creates the impression that we know a lot more than is really the case. Listing sources of uncertainty and the direction of bias is a valuable means for expressing what we do not know (Covello and Merkhofer 1993). By stating which sources of uncertainty are the most critical, future research can be directed at filling the data gaps that would have the greatest impact on reducing uncertainties.

In cases where particular environmental parameters have a strong influence on chemical bioavailability, developing WQC as functions (e.g., aluminum criterion as a function of DOC and pH) would reduce uncertainty by quantifying the influence of the most important confounding factors and by reducing reliance on conservative default values. Thus, the difference between the actual and desired levels of protection is narrowed. This approach has been used on occasion in the past, although much more could be done given the current state of knowledge on factors affecting toxicity of metals, nonpolar organic chemicals, and other classes of chemicals.

The final step in the derivation of WQC is often country specific and involves dividing the critical effects level (e.g., EC25 of the most sensitive species, 5th percentile of the species sensitivity distribution) by an uncertainty or AF. Typically, these factors are arbitrary. Uncertainty could be reduced by relying on empirically based factors (Calabrese and Baldwin 1993; Chapman et al. 1998) or by making better use of existing information. For example, instead of arbitrarily dividing the 5th percentile species sensitivity distribution by a factor of 2 to convert from the effect level (e.g., LC50, LOEC) to a no-effect level, one could estimate the slopes of the concentration–response relationships for sensitive species and derive a more empirically based factor.

Interval analysis is a simple method for uncertainty propagation through mathematical expressions. With interval analysis, it may be feasible to estimate the bounds in which a national WQC could lie, given the desired level of protection. This approach would be particularly useful for criteria that require numerous inputs in the derivation procedure (e.g., soil and sediment-quality criteria [SQC]). Interval analysis is applicable no matter what the nature or source of the uncertainty. Given that the input intervals are accurate reflections of what is known about their respective variables, and that the equation used to derive the criterion is an appropriate model of the way these variables interact, the resulting interval could not be any narrower and yet still be guaranteed to enclose the true answer.

Each of the methods described above for characterizing and reducing uncertainty with Type 1 criteria also can be used with Type 2 site-specific WQC. Use of resident species and site-specific parameters in the derivation of Type 2 criteria further reduces uncertainty and narrows the gap between the actual and desired levels of protection (Figure 4-4).

Type 3 criteria

There are a number of methods that can be used to characterize and reduce uncertainty for Type 3 criteria. Some of these methods require intensive efforts and extensive information. As such, they are not feasible for the majority of chemicals for which WQC are currently derived. Several of the possible methods are briefly described below.

A weight-of-evidence analysis is a weighted evaluation of all lines of evidence that can be used to justify or modify a WQC. The purpose of this approach is to account for, and make logical use of, all the data available so that the final result of the derivation is justified by the preponderance of evidence. Suppose that we have a Type 2 site-specific criteria. If the evidence from a number of in-stream toxicity tests and field biomonitoring studies indicates no observable effects at concentrations well above the Type 2 criterion value, then a weight-of-evidence approach should be invoked to appropriately adjust the value. This would help to close the gap between the actual and desired levels of protection. An example of a simple weight-of-evidence analysis is shown in Table 4-1.

Table 4-1 Hypothetical weight of evidence used to develop Type 3 criteria

Evidence	Result	Details
Biological surveys	No effects at levels up to 2X Type 2 WQC.	Fish community productivity and species richness both high and comparable to pristine conditions in all but most contaminated (2X WQC) site.
Ambient toxicity tests	No effects observed. Chemical concentrations exceeded Type 2 WQC by factor of 1.6.	High lethality to fathead minnow larvae in one test, but result discounted because dissolved oxygen levels were below 1 mg/L.
Whole effluent toxicity (WET) tests	Minor mortality effects in acute WET tests in which chemical concentrations were 4- to 8-fold higher than Type 2 WQC.	Mortality was 12% in WET test with highest chemical concentration.
Type 3 WQC	*Revise Type 2 WQC upward 1.6- to 2-fold higher.*	*Continue biomonitoring programs to ensure Type 3 WQC is protective.*

With the information available in Type 3, it is possible to quantify the uncertainty about the WQC. First-order error analysis, Monte Carlo simulation, probability bounds analysis, fuzzy arithmetic, and a variety of other techniques can be used to propagate uncertainties, given the criterion equation and input distributions. The result will be an output distribution showing the possible values for the Type 3 criteria and their respective probabilities, given the desired level of protection.

The most commonly used technique for uncertainty analysis is Monte Carlo analysis. In a Monte Carlo analysis, point estimates in a model equation are replaced with probability distributions, samples are randomly taken from each distribution, and the results are tallied, usually in the form of a probability density function or cumulative density function.

Monte Carlo techniques excel when large quantities of data and theory exist to properly specify the model equation and the input probability distributions. Difficulties arise when there is insufficient information to specify input distributions or the relationships among them (Smith et al. 1992; Ferson and Long 1994; Bukowski et al. 1995). In such cases, other techniques including probability bounds analysis or fuzzy arithmetic should be considered because the results of such analyses do not depend on knowledge about the covariance relationships among input variables (Ferson and Kuhn 1994; Ferson and Long 1994). Other numerical methods for uncertainty propagation or updating include 1) various Bayesian methods including Bayesian Monte Carlo (Burmaster and Anderson 1994) and 2) hybrid arithmetic (Cooper et al. 1996).

The choice of the appropriate method depends on the complexity of the analysis, the available information, and the expertise of the assessor. The following are general points of guidance for choosing an appropriate uncertainty propagation method (Landis et al. 1998):

- In early tiers, or when little information is available, interval analysis can be used to generate uncertainty bounds (analogous to best-case and worst-case scenarios).

- For simple additive and log-transformed multiplicative models where the variables are independent, an analytical approach such as first-order error propagation is a simple and effective tool for propagating uncertainty.

- For more complex models, Monte Carlo simulation methods may be the most appropriate uncertainty propagation tool, but only if sufficient information exists to adequately characterize the input distributions and the relationships among them.

- When information is limited, other methods with less restrictive requirements should be considered (e.g., probability bounds analysis, fuzzy arithmetic).

The principles of best practice developed by Burmaster and Anderson (1994) for Monte Carlo analysis should be followed when conducting any type of quantitative uncertainty analysis. Uncertainty analysis makes clear what is known and what is not a huge advantage over the arbitrary use of conservative assumptions and safety factors. Thus, uncertainty analysis provides an objective and transparent means of comparing assumptions, models, and data used in the WQC derivation process.

Threatened and endangered species and organisms of special concern

For WQC, the issue of the actual level of protection is especially important for threatened and endangered species because of the potential ecological consequences and the statutory requirements that apply to protection of these species in many jurisdictions.

The USEPA approval of state and tribal water-quality standards is a federal action subject to the consultation requirements in the U.S. Endangered Species Act (ESA). Because there are relatively few toxicity data for threatened and endangered aquatic species, there is uncertainty about the protectiveness of the WQC for these species. Without species-specific data and because of the potential consequences of being wrong, an argument often is made that the threatened and endangered species are sensitive. The principal reasons given for listing of threatened and endangered aquatic species are habitat loss and introduction of exotic species, and there is no toxicological basis for expecting that threatened and endangered species are more sensitive to toxicants than are the surrogates normally used in toxicity testing. In the limited situations where this has been investigated in side-by-side testing of surrogates and threatened and endangered species (cultured organisms), the surrogates have generally been good indicators of the sensitivity of the tested threatened and endangered aquatic species (Dwyer et al. 1995). Nevertheless, this remains an area of uncertainty that warrants further investigation.

In addition to direct toxic effects, it is important to consider potential indirect effects on threatened and endangered species. These can include effects such as dietary exposure through consumption of contaminated prey, or indirect effects related to effects on other organisms critical in the life cycle of the threatened or endangered species (e.g., the larval stage of certain freshwater endangered mussels relies on specific fish species to serve as their host during development). Furthermore, the ESA applies to individuals, and therefore, it may not be sufficient to apply the "protect most of the species most of the time" approach inherent in the national criteria.

Characterization of risk to threatened and endangered species should be an important consideration in the derivation of the national WQC or in any site-specific adjustments to the national criteria. An approach to addressing this issue would be to identify appropriate surrogate species and use or develop toxicity data for those species in deriving the national WQC. There may be other, perhaps more efficient, ways to get at this relative sensitivity issue as well (e.g., ensuring that the commonly used surrogate test organisms adequately represent the threatened and endangered species and species of concern). However it is done, the current uncertainty associated with the protectiveness of the national WQC relative to threatened and endangered species needs to be addressed.

Communicating WQC

Experience indicates that published numerical WQC and standards are generally adopted and used without more than a superficial understanding or appreciation of the information incorporated in the data and procedures used to derive WQC. This lack of understanding is relevant to the regulators, as well as to other stakeholders in the process (politicians, industry, and public). Communicating technical information such as WQC is not an easy task, especially if the result of the WQC assessment is contrary to conventional wisdom or to the interests of certain stakeholder groups. Written communications must be understood by the technical community as well as the lay public—requiring extra effort and unique writing skills. Good communication of technical information can be achieved if the following are addressed:

- Know the target audience through use of marketing techniques, such as surveys, interviews, and focus group sessions, to fully understand the stakeholder perception of technical issues and risk.

- Use an iterative process and incorporate a series of exchanges and careful attention to the feedback to design the message.

- Ensure the entire process is transparent. This can be enhanced through the use of appropriate presentation techniques. Appropriate graphical and pictorial formats for presenting complex technical information are often helpful.

- Never appear indifferent to public perceptions of risk. No matter how absurd the statement appears, it should be taken at face value, as a legitimate concern, and addressed as well as possible. Validate the reality of the statement or question, and then proceed to address it.

In the writing of WQC documents, particular attention should be paid to the following:

- Use appropriate graphics to explain relationships in the data.

- Use clear and understandable language with a minimum of technical jargon, acronyms, and complex compilations of data in tables.

- Use explanatory text boxes to flag important issues and interpretations of the data.

- Relate the WQC to other environmental management issues and decision-making.

- Use simple and appropriate analogies to explain complex and potentially controversial issues.

References

Allen TFH, Starr TB. 1982. Hierarchy. Perspectives for ecological complexity. Chicago IL, USA: The University of Chicago Press.

Arthur JW. 1988. Application of laboratory derived criteria to an outdoor stream ecosystem. *Int J Environ Stud* 32:97-110.

Barnthouse LW, Suter II GW, Rosen AE. 1988. Inferring population-level significance from individual-level effects: An extrapolation from fisheries science to ecotoxicology. In: Suter II GW, Lewis M, editors. Aquatic toxicology and hazard assessment. 11th symposium. Philadelphia PA, USA: ASTM. p 289-300.

Baskin Y. 1994. Ecosystem function of biodiversity. *Bioscience* 44:657-660.

Bukowski J, Korn L, Wartenburg D. 1995. Correlated inputs in quantitative risk assessment: The effects of distributional shape. *Risk Anal* 15:215-219.

Burmaster DE, Anderson PD. 1994. Principles of good practice for the use of Monte Carlo techniques in human health and ecological risk assessment. *Risk Anal* 14:477-481.

Calabrese EJ, Baldwin LA. 1993. Performing ecological risk assessments. Chelsea MI, USA: Lewis Publishers.

Caswell H. 1996. Demography meets ecotoxicology: Untangling the population level effects of toxic substances. In: Newman MC, Jagoe CH, editors. Ecotoxicology: A hierarchical treatment. Boca Raton FL, USA: Lewis Publishers. p 255-292.

Chapman PM, Fairbrother A, Brown D. 1998. A critical evaluation of safety (uncertainty) factors for ecological risk assessment. *Environ Toxicol Chem* 17:99-108.

Cooper JA, Ferson S, Ginzburg LR. 1996. Hybrid processing of stochastic and subjective uncertainty data. *Risk Anal* 16:785-791.

Covello VT, Merkhofer MW. 1993. Risk assessment methods: Approaches for assessing health and environmental risks. New York NY, USA: Plenum Press. 309 p.

Detenbeck NE, DeVore PW, Niemi GJ, Lima A. 1992. Recovery of temperate-stream fish communities from disturbance: A review of case studies and synthesis of theory. *Environ Manag* 16:33-53.

Dwyer JF, Sappington LC, Buckler DR, Jones SB. 1995. Use of surrogate species in assessing contaminant risk to endangered and threatened species. Gulf Breeze FL, Washington DC, USA: USEPA, Office of Research and Development. EPA-6-R-96-029.

Farmer D, Hill IR, Maund SJ. 1995. A comparison of the fate and effects of two pyrethroid insecticides (lambda-cyhalothrin and cypermethrin) in pond mesocosms. *Ecotoxicology* 4:219-244.

Ferson S, Kuhn R. 1994. Uncertainty analysis with fuzzy arithmetic. Risk calc user manual. Setauket NY, USA: Applied Biomathematics.

Ferson S, Long TF. 1994. Conservative uncertainty propagation in environmental risk assessments. In: Hughes JS, Biddinger GR, Mones E, editors. Environmental toxicology and risk assessment. 3rd volume. Philadelphia PA, USA: ASTM. STP 1218.

Geckler JR, Hornig WB, Neiheisel TM, Pickering QH, Robinson EL, Stephan CE. 1976. Validity of laboratory tests for predicting copper toxicity in streams. Duluth MN, USA: USEPA, Environmental Research Laboratory Duluth. EPA-600-3-76-116. 192 p.

Giddings JM, Biever RC, Annunziato MF, Hosmer AJ. 1996. Effects of diazinon on large outdoor pond microcosms. *Environ Toxicol Chem* 15:618-629.

Giddings J, Hall Jr L, Solomon K, Adams W, Vogel D, Davis L, Smith R. 1997. An ecological risk assessment of diazinon in the Sacramento and San Joaquin River Basins. Greensboro NC, USA: Novartis Crop Protection. Technical Report 11/97.

Hansen DJ. 1989. Status of the development of water quality criteria and advisories. Water quality standards for 21st century. Narragansett RI, USA: USEPA, Environmental Research Laboratory. p 163-169.

Huggett RJ, Kimerle RA, Mehrle Jr PM, Bergman HL. 1992. Biomarkers. Biochemical, physiological, and histological markers of anthropogenic stress. Boca Raton FL, USA: Lewis Publishers. 347 p.

Landis WG, Matthews GB, Matthews RA, Sergeant A. 1994. Application of multivariate techniques to endpoint determination, selection and evaluation in ecological risk assessment. *Environ Toxicol Chem* 12:1917-1927.

Landis WG, Matthews RA, Matthews GB. 1996. The layered and historical nature of ecological systems and the risk assessment of pesticides. *Environ Toxicol Chem* 15:432-440.

Landis WG, Moore DRJ, Norton S. 1998. Ecological risk assessment: Looking in, looking out. In: Douben PET, editor. Pollution risk assessment and management: A structured approach. Chichester, UK: John Wiley and Sons, Ltd. p 273-309.

Matthews GB, Matthews RA, Ehinger WJ. 1991. Mathematical analysis of temporal and spatial trends in the benthic macroinvertebrate communities of a small stream. *Can J Fish Aquat Sci* 48:2184-90.

Moore DRJ. 1998. The ecological component of ecological risk assessment: Lessons from a field experiment. *Hum Ecol Risk Assess* 3(5):1-21.

Mount DI, Norberg-King TJ, editors. 1985. Validity of effluent and ambient toxicity tests for predicting biological impact, Scippo Creek, Circleville, Ohio. Duluth MN, Washington DC, USA: USEPA, Office of Research and Development. EPA-600-3-85-044.

Mount DI, Norberg-King T, editors. 1986. Validity of effluent and ambient toxicity tests for predicting biological impact, Kanawha River, Charleston, West Virginia. Duluth MN, Washington DC, USA: USEPA, Office of Research and Development. EPA-600-3-86-006.

Mount DI, Norberg-King T, Steen AE, editors. 1986. Validity of effluent and ambient toxicity tests for predicting biological impact, Naugatuck River, Waterbury, Connecticut. Duluth MN, Washington DC, USA: USEPA, Office of Research and Development. EPA-600-8-86-005.

Mount DI, Steen AE, Norberg-King T, editors. 1985. Validity of effluent and ambient toxicity tests for predicting biological impact, Five Mile Creek, Birmingham, Alabama. Duluth MN, Washington DC, USA: USEPA, Office of Research and Development. EPA-600-8-85-015.

Mount DI, Steen AE, Norberg-King T, editors. 1986a. Validity of effluent and ambient toxicity tests for predicting biological impact, Back River, Baltimore Harbor, Maryland. Duluth MN, Washington DC, USA: USEPA, Office of Research and Development. EPA-600-8-86-001.

Mount DI, Steen AE, Norberg-King T, editors. 1986b. Validity of effluent and ambient toxicity tests for predicting biological impact, Ohio River, Wheeling, West Virginia. Duluth MN, Washington DC, USA: USEPA, Office of Research and Development. EPA-600-3-85-071.

Mount D, Thomas N, Barbour M, Norberg T, Roush T, Brandes R. 1984. Effluent and ambient toxicity testing and instream community response on the Ottawa River, Lima, Ohio. Duluth MN, Washington DC, USA: USEPA, Permits Division, and Office of Research and Development. EPA-600-3-84-080.

Norberg-King TJ, Mount DI. 1986. Validity of effluent and ambient toxicity tests for predicting biological impact, Skeleton Creek, Enid, Oklahoma. Duluth MN, USA; Washington DC, USA: USEPA, Office of Research and Development. EPA-600-8 86-002.

O'Neill RV, DeAngelis DL, Waide JB, Allen TFH. 1986. A hierarchical concept of ecosystems. Princeton NJ, USA: Princeton University Press.

Schindler DW. 1996. Ecosystems and ecotoxicology: A personal perspective. In: Newman MC, Jagoe CH, editors. Ecotoxicology: A hierarchical treatment. Boca Raton FL, USA: Lewis Publishers. p 371-398.

Smith AE, Ryan PB, Evans JS. 1992. The effect of neglecting correlations when propagating uncertainty and estimating the population distribution of risk. *Risk Anal* 12:467-474.

Solomon KR, Baker DB, Richards RP, Dixon KR, Klaine SJ, La Point TW, Kendall RJ, Weisskopf CP, Giddings JM, Giesy JP, Hall Jr LW, Williams WM. 1996. Ecological risk assessment of atrazine in North American surface waters. *Environ Toxicol Chem* 15:31-76.

Suter III GW. 1993. New concepts in the ecological aspects of stress: The problem of extrapolation. *Sci Total Environ*, Suppl 93:63-76.

Suter II GW, Barnthouse LW, Breck JE, Gardner RH, O'Neill RV. 1985. Extrapolating from the laboratory to the field: How uncertain are you? In: Cardwell RD, Purdy R, Bahner RC, editors. Aquatic toxicology and hazard assessment. 7th symposium. Philadelphia PA, USA: ASTM. STP 854. p 400-413.

Suter II GW, Barnthouse LW, Bartell SM, Mill T, Mackay D, Paterson S. 1993. Ecological risk assessment. Boca Raton FL, USA: Lewis Publishers.

Tikkanen P, Laasonen P, Muotka T, Huhta A, Kuusela K. 1994. Short-term recovery of benthos following disturbance from stream habitat rehabilitation. *Hydrobiologica* 273:121-130.

Versteeg DJ, Belanger SE, Carr GJ. 1999. Understanding single species and model ecosystem sensitivity: A data based comparison. *Environ Toxicol Chem* 18:1329-1346.

Walker B. 1992. Biodiversity and ecological redundancy. *Conserv Biol* 6:18-23.

Walker B. 1995. Conserving biological diversity through ecosystem resilience. *Conserv Biol* 9:747-752.

Wallace JB. 1990. Recovery of lotic macroinvertebrate communities from disturbance. *Environ Manag* 14:605-620.

Whiles MR, Wallace JB. 1992. First-year benthic recovery of a headwater stream following a 3-year insecticide-induced disturbance. *Freshw Biol* 28:81-91.

Whiles MR, Wallace JB. 1995. Macroinvertebrate production in a headwater stream during recovery from anthropogenic disturbance and hydrologic extremes. *Can J Fish Aquat Sci* 52:2402-2422.

Yount JD, Niemi GJ. 1990. Recovery of lotic communities and ecosystems from disturbance—A narrative review of case studies. *Environ Manag* 14:547-569.

Review of Current Approaches and Future Directions in National Ambient Water-Quality Criteria for Aquatic Life Protection: Canada, Netherlands, New Zealand, USA

Robert A. Kent (Canada)
Trudie Crommentuijn (Netherlands)
Christopher W. Hickey (New Zealand)
F. James Keating Jr. (USA)

Approaches:
Canada, Netherlands, New Zealand, USA

This appendix presents a review and comparative evaluation of national ambient water and SQC from several international jurisdictions. The countries considered are Canada, Netherlands, New Zealand, and USA. Each country's review comprises 3 sections: 1) a description of terminology and legal framework, 2) a description of the criterion derivation methodology, and 3) a discussion of new developments, topic areas, and future directions (related to development and implementation). This is followed by a comparison of the different approaches, an outline of areas of commonality and divergence, and a presentation of the respective rationale (scientific, policy, arbitrary, etc.) for such differences. A selected sample of criteria is presented in a comparative fashion as well.

Reevaluation of the State of the Science for Water-Quality Criteria Development. Mary C. Reiley et al., editors.
©2003 Society of Environmental Toxicology and Chemistry (SETAC). ISBN 1-880611-30-9

Canadian National Water-Quality Guidelines for the Protection of Aquatic Life: Current Status and Future Directions

Robert A. Kent, Pierre-Yves Caux, Susan L. Roe
(National Guidelines and Standards Office, Environment Canada, Hull, Quebec, Canada)

Terminology and Legal Framework

Canadian environmental-quality guidelines (EQGs) are nationally endorsed, science-based goals for the quality of atmospheric, aquatic, and terrestrial ecosystems. EQGs are defined as numerical concentrations or narrative statements that are recommended as levels that should result in negligible risk to biota, their functions, or any interactions that are integral to sustaining the health of ecosystems and the designated resource uses they support. Canadian EQGs are recommended for parameters of national concern that are found in the ambient environment. As national benchmarks or indicators of environmental quality, Canadian EQGs are intended to protect, sustain, and enhance the quality of the Canadian environment and its many beneficial uses.

The following discussion provides an overview of the history, current status, and future directions of Canadian water-quality guidelines (CWQGs) (CCREM 1987) and their relationship to other EQGs (i.e., sediment, tissue, and soil).

CWQGs: History

In 1987, Environment Canada, through the auspices of a federal-provincial task group under the Canadian Council of Ministers of the Environment (CCME; formerly CCREM), developed the first CWQGs. These guidelines included those for the protection of freshwater life, agricultural water uses for irrigation and livestock, raw water for drinking water supply, recreational water quality and aesthetics, and industrial water supplies (CCREM 1987). This publication represented the first time that national, science-based guidelines were developed collaboratively amongst provincial, territorial, and federal jurisdictions.

EQGs: Current status

It has been recognized that, in addition to those for water, guidelines for other environmental media are required to adequately protect the ecosystem. Thus, Environment Canada, in cooperation with the Canadian provinces and territories, develops sediment, soil, and tissue residue quality guidelines to address the need to protect all components of the ecosystem in a more holistic manner. Like CWQGs, sediment quality guidelines are intended to protect all forms of aquatic life, whereas soil-quality guidelines are derived to protect both human and environmental health with respect to different land uses. Tissue residue guidelines are maximum allowable concentrations of contaminants in the tissue of aquatic biota that are not expected to result in adverse effects in the wildlife that consume them. These guidelines, together with guidelines for air quality, community water supplies, and recreational and aesthetic water uses, were published in 1999 in a document entitled "Canadian Environmental Quality Guidelines." This document synthesizes EQG activities in Canada over the past decade and provides a comprehensive and practical reference guide for environmental resource managers across Canada. The availability of EQGs for a number of ecosystem components will facilitate the comprehensive management of aquatic and terrestrial ecosystems and will provide a consistent and common scientific basis for making effective decisions regarding the protection of environmental quality in Canada.

Guiding Principles for the Development of Canadian EQGs

Since the release of CWQGs (CCREM 1987), science-based guideline derivation procedures have been established and approved nationally for specific media and resource uses. These procedures have been documented as national, scientific protocols (CCREM 1987; Health Canada 1989; CCME 1991, 1993, 1995, 1996, 1998; WGAQOG 1996). The use of national protocols ensures consistency, transparency, and scientific defensibility in the guideline development process.

Regardless of the resource uses to be protected, guideline development for individual substances is founded on the same set of guiding principles and follows a consistent process, although specific elements of these protocols may necessarily differ. Three guiding principles are fundamental to the development and implementation of Canadian EQGs:

1) EQGs embody a national goal for environmental quality of no observable adverse effects on atmospheric, aquatic, and terrestrial ecosystems over the long term;

2) EQGs are developed for major atmospheric, terrestrial, and aquatic resource uses in Canada; and

3) EQGs are generic recommendations that are based on the most current scientific information (i.e., they do not directly consider site-specific or management factors that may influence their implementation).

Environmental quality guidelines should not be regarded as blanket values for national environmental quality. Variations in environmental conditions across Canada will affect environmental quality in different ways. Therefore, the users of EQGs may need to consider local conditions and other supporting information (e.g., site-specific background concentrations of naturally occurring substances) during the implementation of EQGs. Science-based, site-specific criteria, guidelines, objectives, or standards may therefore differ from the recommended Canadian EQGs. For ecosystems of superior quality, impairment to guideline concentrations is not advocated.

CWQG derivation protocol

The protocol for the derivation of WQGs for the protection of aquatic life, originally published in 1991, provides a consistent, scientifically defensible approach. CWQGs "are set at such values as to protect all forms of aquatic life and all aspects of the aquatic life cycles." The goal is to protect all life stages during an indefinite exposure to water. Whether this goal can be realized is a water management issue and does not affect the guideline derivation procedure. For most water-quality variables, a single maximum value based on a long-term no-effect concentration, which is not to be exceeded, is recommended. In setting the guideline, all components of the aquatic ecosystem (e.g., algae, macrophytes, invertebrates, fishes) for which data exist are considered. Moreover, a vast suite of parameter-specific information is investigated, including physical and chemical properties, fate and persistence, and environmental levels. Toxicity studies that examine ecologically relevant endpoints (e.g., growth or survival) from exposure to contaminated water are compiled and evaluated according to scientific standards. Moreover, minimum data requirements have been set; where data are lacking, an interim guideline is preferred over no guideline.

Canadian WQGs are calculated separately for freshwater and marine environments, but in both cases are derived preferably from the lowest-observed-effects level (LOEL) from a chronic study using a nonlethal endpoint for the most sensitive life stage of the most sensitive aquatic species investigated. The LOEL, which must be statistically significant, is multiplied by a safety factor of 0.1 to arrive at the guideline value

$$\text{CWQG} = \text{LOEL} \times 0.1 \tag{A-1}.$$

The safety factor of 10 has been chosen to account for differences in sensitivity to a chemical variable due to differences in species, laboratory versus field conditions, and test endpoints. The guideline usually applies to the total amount of contaminant in an unfiltered water sample.

In cases where chronic data are lacking, guidelines may be derived from acute studies by converting short-term median lethal or median effective concentrations (LC50, EC50) to long-term, no-effect concentration. ACRs can be used to convert the median lethal results of a short-term study to an estimated long-term, no-effect concentration. An ACR is calculated by dividing an LC50 or EC50 by the NOEL from a chronic exposure test for the same species (i.e., LC50 / NOEL):

$$\text{ACR} = \text{LC50 or EC50} \div \text{NOEL} \tag{A-2}.$$

If ACRs are not available, the LC50 or EC50 value may be multiplied by a universal AF; the AFs for persistent and nonpersistent variables are 0.01 and 0.05, respectively.

Principal applications and uses of Canadian EQGs

Environmental quality guidelines have a number of functional uses within various environmental assessment and management strategies. Potential applications of EQGs include the scientific basis for the development of environmental regulations, site-specific criteria, guidelines, objectives, or standards including Canada-wide standards under the Canada-wide Accord on Environmental Harmonization:

- indicators for state of the environment reporting;
- science-based goals or performance indicators for regional, national, or international management strategies for toxic substances, including assessment and remediation of contaminated sites;
- scientific tools for assessing risks associated with existing concentrations of persistent, bioaccumulative, and toxic substances in the ambient environment and for tracking progress towards their virtual elimination;
- tools to evaluate the effectiveness of point-source controls; and
- national benchmarks to assess potential or actual impairment of socially relevant resource uses.

Future Directions

Since the publication of the protocol, several issues regarding the derivation procedure for CWQGs have been raised.

Safety factors

Margins of safety will continue to be based on science, socioeconomic, and political considerations. Criticism of their use stems from the notion that they are arbitrary and not science based. Consequently, there may be a perception that environmental risk is overestimated, leading to undue costs to society. Although one can argue that originally margins of safety were based on science, 3 to 4 decades of new knowledge have not influenced their derivation. Margins of safety have become a misunderstood, poorly communicated catchall quantity to account for the natural variability and our lack of knowledge (uncertainty) about a system under study. They often persist to be coarse multipliers of 10 (i.e., 10, 100, 1000) by which we divide calculated quantities such as low threshold toxicity estimations (e.g., LOECs, EC20).

We are developing and proposing a new approach to the setting of margins of safety. It begins with teasing out and communicating what is known (quantifiable) and what is unknown (not quantifiable) about a system. An emphasis is placed on regression analysis because it provides approaches that use whole data envelopes.

Metal-specific issues

Since the publication of the guideline development protocol for CWQG in 1991, environmental science has made progress in many areas, and as a result, certain aspects of the protocol have been identified as being in need of revision, particularly with respect to metals. Issues such as differing toxicity for varying chemical species of a metal, potential conversion of one metal species into another due to changing environmental conditions, influence of the water chemistry (i.e., pH, hardness, counter ions, organic matter, complexing agents, redox, etc.) on the bioavailability and, ultimately, the toxicity of the metal, essentiality of certain metals, and potential for recommending a national guideline that could be below the natural environmental background concentration at particular sites (i.e., national versus site-specific) were highlighted and are targeted for inclusion in an upcoming addendum to the guideline derivation protocol.

Site-specific application of WQGs in Canada

While the CWQGs are intended to protect the designated uses of aquatic ecosystems, it is possible that the guidelines are over- or underprotective at sites with unique conditions. For example, the most sensitive species that occurs at a site may be more or less sensitive than the most sensitive species represented in the toxicological data set that was used to derive the guidelines. Similarly, a substance may be more or less toxic in site water (i.e., due to factors such as pH, water hardness, complexing agents, etc.) than it is under the range of conditions that is represented in the toxicological data set. In part, concerns related to the applicability of the guidelines can be addressed by considering information on site characteristics in the development of site-specific water quality objectives. Moreover, where guidelines for different media (i.e., water, soil, tissue) exist for a single substance, guideline values should reflect the substance's fate, behavior, and persistence among the different media.

Current Approaches for National Ambient Water-Quality Criteria for Aquatic Life Protection in the Netherlands

Trudie Crommentuijn
(Dutch National Institute for Public Health and the Environment, Centre for Substances and Risk Assessment, Bilthoven, Netherlands)

This section presents a summary of the methodology to derive national AWQC and SQC in the Netherlands. In the first section, the terminology and framework are presented. In the second, the methodology is presented. The last section discusses some new developments and future directions.

Terminology and Legal Framework

In the Netherlands, the terms "ecotoxicological risk limit" (ERL) and "environmental-quality standard" (EQS) are used. ERLs are scientifically underpinned values and present a concentration below which, with a certain probability, a degree of effects on ecosystems is expected. A risk limit is based on scientific information only (VROM 1989, 1994). EQSs are set by the Ministry of Housing, Physical Planning, and the Environment (VROM) and may sometimes take into consideration other factors (e.g., economic and social) in addition to the risk limit. The different levels of ERLs used in the Netherlands, ecotoxicological serious soil contamination concentration/human toxicological serious soil contamination concentration (ECOTOX SCC/HUMTOX SCC), MPC, and negligible concentration (NC), and the different corresponding EQS, intervention value, MPC, and target value, are presented in Table A-1. Intervention values are derived for soil, groundwater, and sediment. MPCs and target values are derived for water, soil, and sediment.

Table A-1 ERLs and EQSs in the Netherlands[a]

ERLs	EQSs
ECOTOX SCC or ECOTOX SCC and HUMTOX SCC[b]	Intervention value for soil clean-up and sediment
MPC	MPC
NC	Target value

[a]VROM 1994.
[b]HUMTOX SCC not dealt with in this paper.

Depending on the level of the EQS, different actions may be needed once a value is exceeded. The type of action needed is also dependent on the compartment considered. For water, MPCs and target values are derived. These values are compared with data from measurement programs. If values are exceeded, these substances become so-called "priority substances" and are indicated as such every 4 years when the National Policy Document on Water Management is published by the Ministry of Traffic and Water Management. This may result in either a more sophisticated "site-specific" risk assessment for certain locations or the decision that a program is needed to reduce the discharges on a national scale, depending on the degree of exceedance. If the target value is exceeded, no defined actions are needed, but the substance in question will be maintained in monitoring programs to check whether or not concentrations are rising. The level of the target values is considered a reference value, a level that environmental policy wants to reach in the long run.

For sediment, besides the MPC and target values, the intervention value is also used. If the MPC and target values are exceeded by concentrations derived from monitoring programs, the same conditions and actions are needed as mentioned above for water. Besides this, the intervention value is used in the monitoring context. When this value is exceeded, sediment cleanup is needed in principle, but not before a more refined site-specific effect assessment can prove that there is a real risk as large as described by the "generic" values. The MPC and intervention values are also used to evaluate the toxicity of sediment removed from harbors. Before it is decided where the sludge should go, a more refined site-specific effect assessment is performed by 1) refining the generic value on the basis of site-specific physicochemical characteristics, and 2) performing a standard set of bioassays to determine whether the sediment is as toxic as judged on the basis of the generic quality standard. On the basis of these results, it is decided where the sludge should go: dumped into the sea off the coast or put away "safely" so that transport to the sea is not possible.

Methodology to Derive Generic ERLs

Figure A-1 presents a schematic outline to derive risk limits, consisting of 4 steps. In each step, decision criteria are described with procedures that must be completed before the next step can be taken.

In the first step, criteria describe what data are needed from the scientific literature, particularly toxicological data for responses at the population level. In general, these are survival, growth, reproduction, and other ecologically relevant parameters. They are expressed commonly as an LC50 or EC50 (short-term tests, duration 4 days or less) or as a NOEC (long-term tests, duration more than 4 days, with the

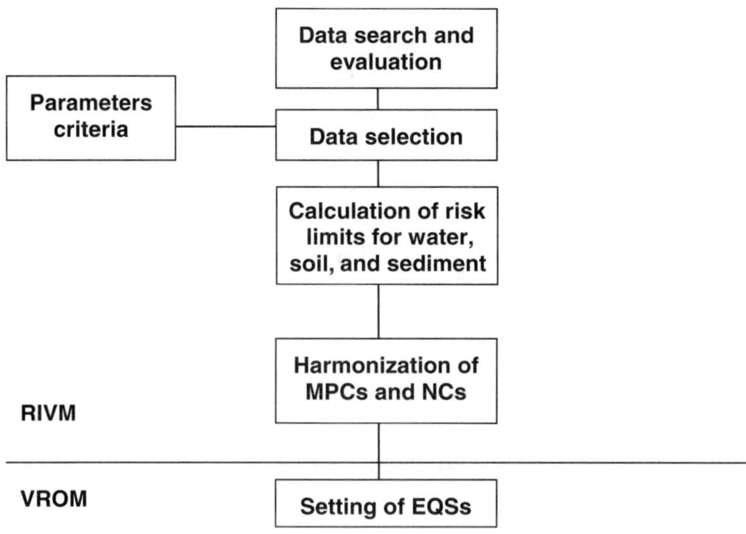

Figure A-1 The process of deriving generic ERLs

exception of microorganism and algal tests for which an NOEC may be derived from short-term experiments). If there are indications that the substance in question has a potential for bioaccumulation, data on the sensitivity of birds and mammals and BCFs for worms and fish are also necessary. Only data passing the quality criteria, as embedded in the Centre for Substances and Risk Assessment Quality System (CSR 1996), are taken into account.

In Step 2, one NOEC value per species has to be selected or calculated. Chronic as well as acute toxicity data are weighed over the species as follows (Slooff 1992):

a) If, for one species, several toxicity data based on the same toxicological endpoint are available, these values are averaged by calculating the geometric mean.

b) If, for one species, several toxicity data based on different toxicological endpoints are available, the lowest value is selected. The lowest value is determined on the basis of the geometric mean, if more than one value for the same parameter is available (see [a] above).

- In some cases, data for effects on different life stages are available. If it becomes evident from these data that a distinct life stage is more sensitive, this result may be used in the extrapolation.

In Step 3, the selected data are used as input in the extrapolation methods. Refined effects assessment or statistical extrapolation is applied if chronic data for more than 4 different taxonomic groups are available (Aldenberg and Slob 1993). A 95% protection level is chosen as the level of protection for the MPC (VROM 1989). The statistical extrapolation method is elaborated into the "added risk approach" to be used for all anthropogenic and naturally occurring substances (Struijs et al. 1997). The ECOTOX SCC is assumed to represent a "serious threat" to the ecosystem and the level of protection is set at 50% (VROM 1989). The NC, in contrast to the MPC and the ECOTOX SCC, is not based on a fraction of species protected, and is derived by dividing the MPC by a factor of 100. This factor is applied to take into account combination toxicity and uncertainties in risk assessment (VROM 1989).

The preliminary effects assessment method is one in which assessment factors are applied to toxicity data, and it is used if fewer than 4 chronic NOECs are available. The size of this factor depends on the number and kind of these toxicity data and is described by Van de Meent et al. (1990) and Crommentuijn et al. (1994) for the MPC and the ECOTOX SCC, respectively.

For substances with a bioaccumulation potential, data on the sensitivity of birds and mammals and BCFs for worms and fish must also be searched for to derive a secondary poisoning risk limit (Everts et al. 1992; Romijn et al. 1992; Jongbloed et al. 1994; van de Plassche 1994). The substances for which this step is considered are organic substances with a log K_{ow} >5 and a molecular weight <600. Metals are considered on a case-by-case basis. The resulting MPC or ECOTOX SCC for birds and mammals is recalculated into a concentration in the soil, using the risk limit for birds and mammals expressed as a concentration in the food and the BF for the food (van de Plassche 1994). ERLs for sediment are derived by applying the equilibrium partition (EP) method, which is also used for harmonization.

In Step 4, harmonization of soil and water risk limits is performed by applying the EP ecotoxicological risk limit method (Pavlou and Weston 1984; Shea 1988; Di Toro et al. 1991). Harmonization of soil and water risk limits with risk limits for air, which are based on human toxicological data, is performed by using multimedia fate models like Simple Box (Van de Meent and De Bruijn 1995).

New Developments

Deriving quality criteria is a compilation of methods and knowledge available at this moment. Because of ongoing efforts to improve effects assessment, it is expected that new methods can be incorporated in the future. In the following new developments, topic areas and future directions (related to development and implementation) being explored by the different countries are presented.

At the moment, risk limits are based on total concentrations. However, it can be expected that the bioavailability in the laboratory is different from the bioavailability in the field. The nature of these differences requires additional investigation. The added risk approach (Struijs et al. 1997) served as the starting point for a project entitled "Risk Assessment of Heavy Metals." In this project, attention is paid to the question of how to determine the bioavailable concentrations of heavy metals to enable a better estimate of the risks imposed by heavy metals in the future (Janssen, Peijnenburg et al. 1997; Janssen, Posthuma et al. 1997; Peijnenburg et al. 1997).

Some metals for which EQSs have been developed are essential elements. Being an essential element means that species may have a minimum requirement for the element in question that supplies its needs, and a maximum above which the element is toxic (Scheinberg 1991). This minimum requirement is necessary because trace metals play an essential role in the metabolism of the organism (Rainbow 1993). In ecotoxicological studies with essential elements, it is theoretically possible to observe effects caused by element limitation instead of toxic effects. Insufficient knowledge is available at the moment to judge whether current EQSs are protecting against deficiency effects.

To "validate" EQSs, quality criteria are compared with effects on more than one species in (semi)field conditions. This method is currently the most applicable to be used in effects assessment procedures. Assessment of the toxic effects of contaminants along a gradient from a point source of pollution (e.g., smelter stack) is one method of associating ecotoxicological risk assessment with toxic effects on population and community parameters in the field (Posthuma 1997). Another method may be the validation of some extrapolation methods with toxicity data derived from multiple-species experiments or mesocosms, as was done by Emans et al. (1993) using aquatic toxicity data.

Australian and New Zealand Environment and Conservation Council Approach to Derivation of Water-Quality Guideline Values for Toxic Contaminants in Aquatic Ecosystems: Heavy Metals[1]

Christopher W. Hickey
(National Institute of Water and Atmospheric Research, New Zealand)

Terminology and Legal Framework

The Australian and New Zealand Environment and Conservation Council (ANZECC) has undertaken a revision of the WQGs for Australia and New Zealand. The previous ANZECC Guidelines (ANZECC 1992) for toxic chemicals followed the Canadian (CCREM 1987) approach, "... to protect all forms of aquatic life and all aspects of the aquatic life cycle... The intention is to protect all life stages during indefinite exposure to the water," and essentially used the Canadian values. The revised ANZECC Guidelines (ANZECC and the Agriculture and Resource Management Council of Australia and New Zealand [ANZECC & ARMCANZ] 2000) use the Dutch and OECD risk-based approach that was recommended for New Zealand (Ministry for the Environment [MfE] 1996) and Australia (Warne 1998), based on reviews of the legislative philosophies and methods of guideline derivation.

This paper presents the results of the application of the risk-based approach to guideline calculation for a selected range of priority metals and discusses their application in relation to deriving site-specific guidelines relevant to the New Zealand legislation. A core set of priority metals was selected for numeric guideline derivation for fresh and marine waters. The 6 priority metals were copper, chromium, arsenic, cadmium, zinc, and lead. The choice of these metals relates to their widespread use in timber preservation (copper, chromium, arsenic), their presence in phosphatic fertilizers and batteries (cadmium), and as components of stormwater runoff (copper, zinc, lead). Data for these metals will be used to illustrate the application of this approach using available databases, to assess the relative sensitivity of organism groups, and to assess the adequacy of the existing databases for derivation of risk-based guidelines from existing chronic data.

---

[1] This section of Appendix 1 summarizes a poster presentation from the workshop. Hickey and Pyle (2001) address the guidelines derivation in greater detail.

The objectives were

- calculation of numeric guidelines for fresh and marine waters for 6 priority metals (copper, chromium, arsenic, cadmium, zinc, and lead),
- categorization of the type of organisms contributing to the numeric guidelines, and
- provision of guidance on the appropriate approach for use with narrative guidelines.

New Zealand legislation

The New Zealand Resource Management Act (RMA) (1991) provides narrative standards for environmental protection. Policy and national guidance is provided by the MfE, with implementation by 16 regional councils. Under the RMA, regional councils can prepare plans that specify the values for which waters are to be managed. The RMA provides some guidance on the choice of values, such as fish spawning or fisheries or aquatic ecosystem protection. The "aquatic ecosystem protection" class is stringent and requires that a discharge shall not be allowed if, after "reasonable mixing," there is any "adverse effect" to the aquatic ecosystem. The more general requirement for no "significant adverse effects" must be met for all discharges. A significant adverse effect is not defined in the RMA. Thus, there is a need to consider derivation of guidelines for environments with differing levels of risk.

Technical Approach

The MfE (1996) suggested that the following principles should be included in a methodology for calculating guidelines values:

- incorporate the precautionary principle, as a key component of sustainable management;
- be able to calculate different levels of protection to suit a particular situation and the values that are to be sustained; and
- use a "transparent" methodology so that the community can understand how a particular guideline value was derived.

The Dutch and OECD risk-based statistical approach meets the principles listed above. It 1) uses NOEC data, so it is inherently precautionary, 2) can calculate different levels of protection, and 3) is transparent and easy for the community to understand conceptually. The risk-based statistical approach of Aldenberg and Slob (1993) and the ETX Model (Aldenberg 1993) was used. The statistical method calculates the 99th and 95th percentile levels of protection with 50% confidence (see Figure A-2 for illustration). The key assumptions used in the Dutch approach are these:

- The toxicity data are distributed log-logistically.
- The ecosystem is adequately protected, provided that a certain percentage of species is protected. The Dutch have chosen a level of 95%; the test species are randomly selected from the ecosystem.
- There are no interactions between species in the ecosystem.
- The NOEC data are the most appropriate data to use to set ambient environmental guidelines.
- The NOEC data for 5 species representing different taxonomic groups constitutes a sufficient dataset.

Information was obtained from USEPA Aquatic Information Retrieval (AQUIRE) database; Australian Ecotox Database; existing Dutch database; published journal reviews; and the Aquatic Sciences and Fisheries Abstracts (ASFA) database.

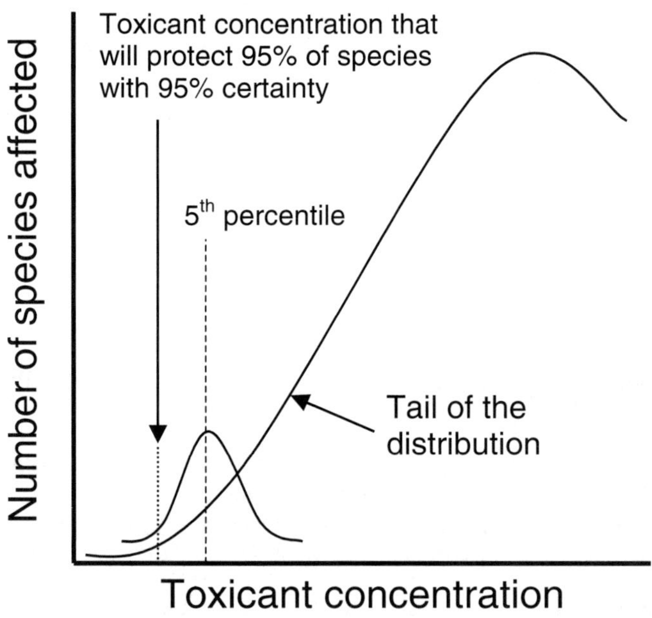

Figure A-2 The Dutch statistical approach, shown in conceptual form. The distribution is assumed to be logistic. The smaller bell-shaped curve shows the confidence interval for the 95th percentile.

Findings

Guidelines were calculated for 17 metals for freshwater species and 10 metals for saltwater organisms (Hickey and Pyle 2001). Comparison of species sensitivity distributions provided insight into the relative sensitivity of various species groups and their ecological susceptibility to that contaminant.

Characteristics of chronic data sets were as follows:

- The range of sensitivity of freshwater biota to metals was 4.5 to 6 orders of magnitude (e.g., Figure A-3 data for copper), with a lower sensitivity range for seawater organisms (3 to 5.5 orders of magnitude).

- The toxicity data for metals (e.g., copper, cadmium, zinc, lead, and CrVI) in freshwater represented a wide range of taxonomic groups, but there were relatively few data for other metals of concern (e.g., CrIII, arsenic).

- Most of the data sets did not have either highly sensitive or insensitive outliers. The relative sensitivity of the various species groups showed marked differences for some metals (e.g., crustaceans and fish for freshwater CrVI, Figure A-4).

- The data sets include very few Australian or New Zealand studies with native species. This largely results from the use of chronic data; the majority of Australasian studies are of acute toxicity.

Guideline values

The ANZECC & ARMCANZ (2000) risk-based guideline derivation procedure provides numeric values corresponding to differing levels of ecosystem protection (e.g., 99%, 95%, 75%). These numeric "trigger values" provide protection levels for application to environments requiring differing levels of protection (e.g., 99% for pristine environments, 95% for general use). The findings were as follows:

- The 99% protection guideline values calculated by the Aldenberg and Slob method were generally lower (by 1.4 to 70 times) than existing ANZECC (1992) guideline values (Table A-2).

- The 95% guideline values ranged from about 3X above (e.g., marine cadmium) and below (e.g., freshwater CrVI) the ANZECC (1992) guideline values.

- The 99% protection guideline values were generally lower than lowest chronic data (e.g., copper, Table A-2), though larger data sets without hardness relationships gave values approximating the 95th percentile of the data set (e.g., CrVI, Figure A-4).

- These calculations showed that 99% protection guideline values would be particularly low and may often be at or below measured background levels for the various metals.

Copper (freshwater)

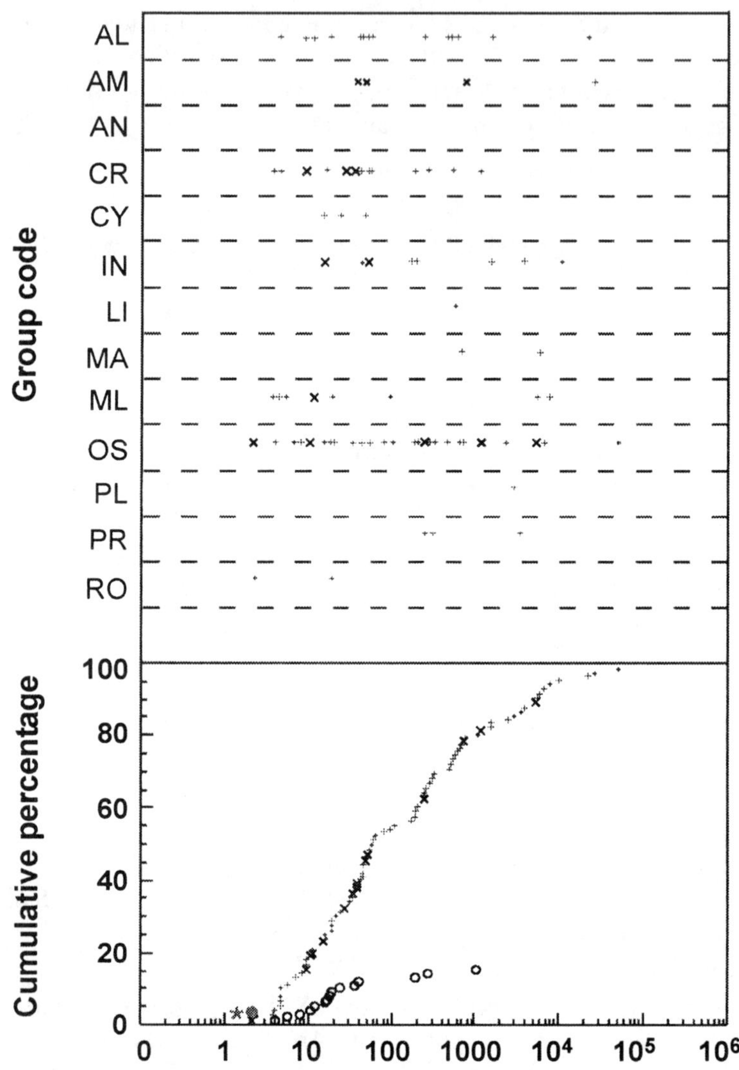

Figure A-3 Summary data for copper in freshwater, showing cumulative frequency plot of organism sensitivity (lower panel) and organism sensitivity by taxonomic group (upper panel).

+ = base data.
x = data with hardness values.
o = hardness corrected data used in ETX calculations.
· = existing ANZECC (1992) guideline.
* = proposed revised guideline (95% protection value).

AL = algae.
AM = amphibians.
AN = annelids.
CR = crustaceans.
IN = insects.
LI = floating macrophytes.
MA = macrophytes.
ML = molluscs.

OS = fish.
PL = planarians.
PR = protozoa.
RO = rotifers.

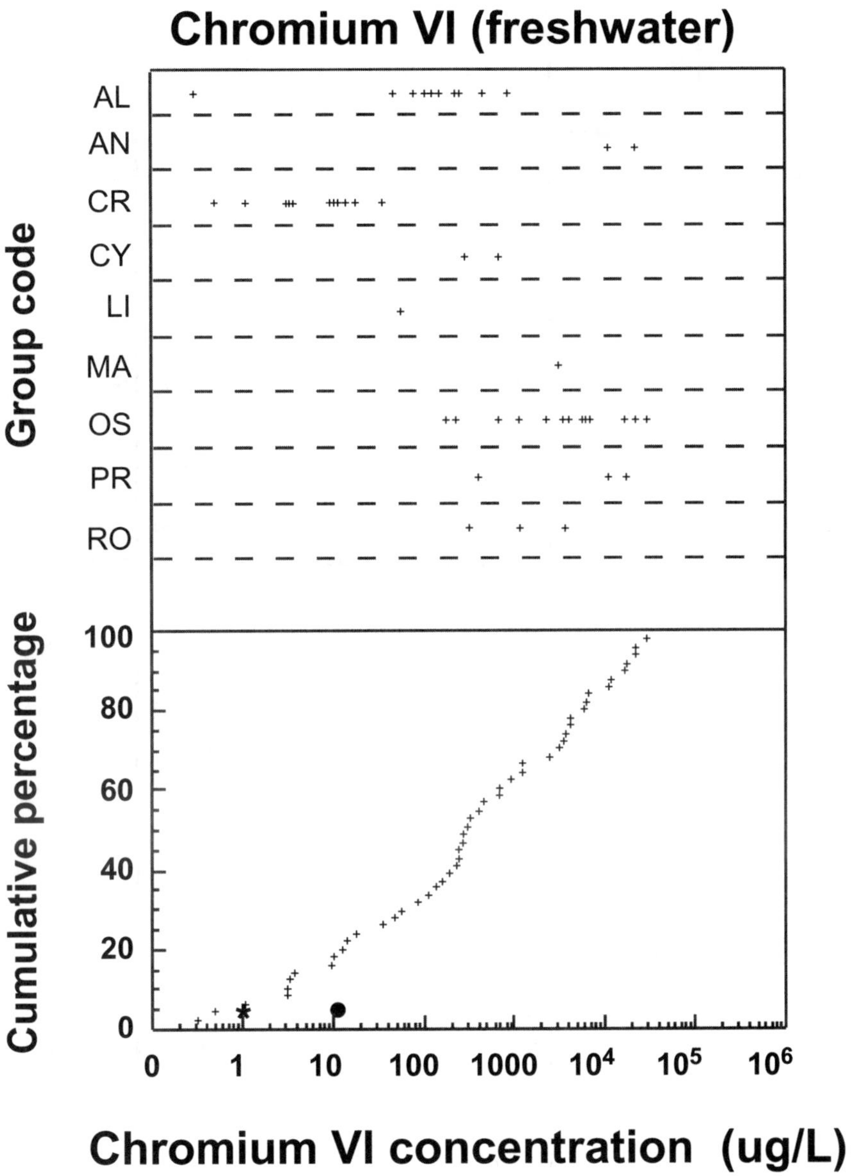

Figure A-4 Summary data for chromium (VI) in freshwater, showing cumulative frequency plot of organism sensitivity (lower panel) and organism sensitivity by taxonomic (upper panel) group.

+ = base data.	AL = algae.	OS = fish.
x = data with hardness values.	AM = amphibians.	PL = planarians.
o = hardness corrected data used	AN = annelids.	PR = protozoa.
in ETX calculations.	CR = crustaceans.	RO = rotifers.
· = existing ANZECC (1992)	IN = insects.	
guideline.	LI = floating macrophytes.	
* = proposed revised guideline	MA = macrophytes.	
(95% protection value).	ML = molluscs.	

Table A-2 Guideline values calculated for a selection of metals to freshwater and marine organisms[a]

	Freshwater				Marine			
Metal	Lowest chronic (µg/L)	Guideline level 99%[b] (µg/L)	Guideline level 95%[b] (µg/L)	Existing ANZECC guideline	Lowest chronic (µg/L)	Guideline level 99%[b] (µg/L)	Guideline level 95%[b] (µg/L)	Existing ANZECC guideline
Copper[c]	0.85	0.60	1.0	2.0–5.0	2	0.49	2.6	5.0
Cadmium[c]	0.15	0.008	0.11	0.2–2.0	0.9	1.4	11	2.0
Zinc[c]	7.9	0.76	6.4	5.0–50	18	2.4	16	50
Lead[c]	0.40	0.30	1.7	1.0–5.0	16	1.4	12	5.0
Arsenic III	40	7.6[d]	60[d]	50	60	0.74[d]	17[d]	50
Arsenic V	10	0.34[d]	6.2[d]	—	25	0.70[d]	9.2[d]	—
Chromium VI	0.51	0.18[d]	2.8	10	0.3	2.9	25	50
Chromium III[c]	6.1	2.3[d]	11	—	469	24	122	—

[a] Modified from Hickey and Pyle 2001.
[b] From ETX software program based on logistic fit.
[c] Freshwater values calculated for a hardness value of 30 mg $(CaCO_3)$/L.
[d] Poor database—low reliability values.

- The 95% guideline values exceeded the lowest chronic data point for several metals, indicating that site-specific consideration of species sensitivity may be required for these metals.

Recommendations

Guideline values were calculated for freshwater and marine organisms using a risk-based statistical method. The statistical method calculates guideline values with different levels of protection and confidence.

The recommendations relative to the New Zealand narrative legislation were
- that the use of the 99% protection guideline be used for protection of waters from "adverse effects," the highest protection level;
- that 95% guideline values are appropriate for protection from "significant adverse effects," the general classification for receiving waters;
- that guideline values apply to soluble metals (i.e., <0.45 μm) above background concentrations; and
- that a review of international and Australasian data sets be required to establish background concentrations of essential elements in "pristine" environments, in order to establish minimum values for essential elements.

Adequacy of data sets

A number of areas were identified to improve the quality and defensibility of the data sets used for guideline derivation. The 3 recommendations were these:

1) Adequacy of data sets
   - Additional testing should focus on underrepresented freshwater aquatic insects and marine fish species.
   - Additional reviews are required for data inclusion in relation to chemical validation and endpoints measured.
   - Appropriate representative regional testing species should be identified and chronic test methods developed for those species.

2) Improvements to statistical method
   - Rigorous testing of the sensitivity of the statistical model (e.g., ETX) for effects of the number of chronic data, and of deviations from the logistic model assumptions in the tail regions, should be undertaken prior to promulgating guideline values (note that the ANZECC 2000 guidelines use a revised Burr III distribution);
   - "Best professional judgment" must be required for some data sets because of limited numbers of data and deviation from model distributions.

Acknowledgments: I am particularly grateful to Lisa Golding for ETX processing, Jóke Baars for literature searches and obtaining publications, Steph Turner for reviewing publications for inclusion; and also to Eric Pyle, MfE and John Chapman and Michael Warne at the Centre for Ecotoxicology, Sydney, Australia for their support in this project.

USA Ambient Water-Quality Criteria

F. James Keating Jr.
(USEPA, Office of Science and Technology, Office of Water, USA)

Terminology and Legal Framework

The objective for ambient water quality in the U.S. is to restore and maintain the chemical, physical, and biological integrity of the nation's waters. The U.S. Clean Water Act addresses water pollution through technology-based and water-quality-based controls. WQC are a component of the water-quality-based approach to pollution control (see Figure A-5).

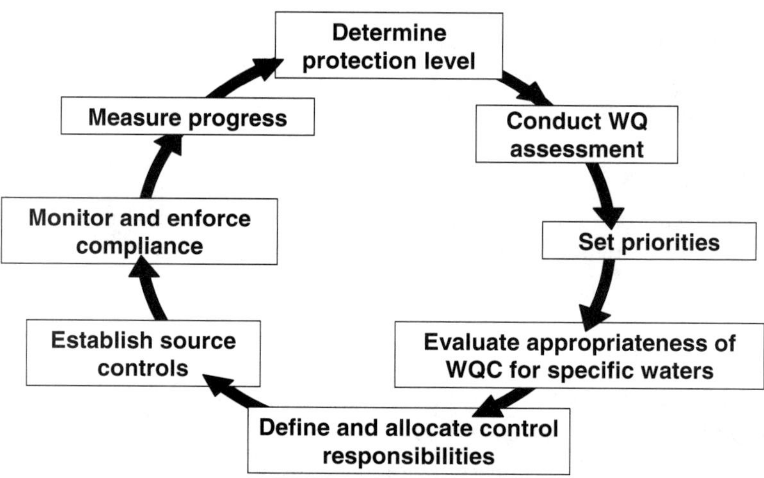

Figure A-5 The water-quality-based approach to pollution control in the U.S.

Step 1 of this process is determining the protection level for a water body. In the U.S., a WQS defines water-quality goals for a water body. A WQS includes a designated use, WQC, an antidegradation policy, and implementation procedures (see Figure A-6).

The Clean Water Act specifies water quality that provides for protection and propagation of fish, shellfish, and wildlife, and for recreation in and on the water. Once adopted, a WQS becomes the basis for legally enforceable control mechanisms. Every 3 years, states submit WQSs for their waters, for review and approval by the USEPA.

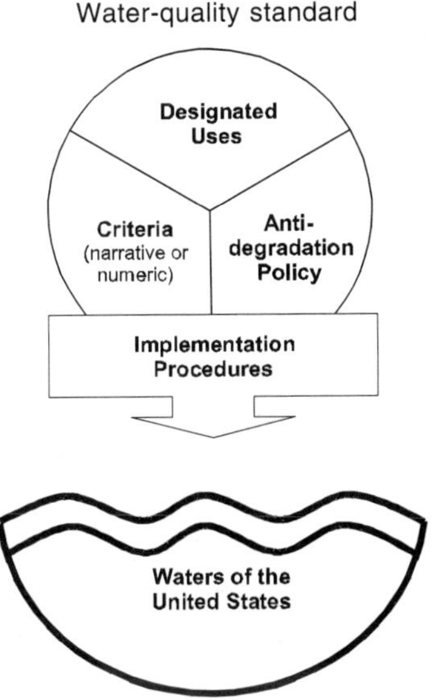

Figure A-6 Components of a WQS

Water-quality criteria are limits on a particular pollutant or on a condition of a water body intended to protect and support a use. WQC can take the form of numeric expressions or narrative statements that describe a desired condition. When criteria are properly selected and met, it is expected that the water quality will protect the designated use. States and tribes are encouraged to adopt numeric criteria for specific pollutants based on published USEPA recommendations, site-specific modifications to USEPA recommended values using established USEPA procedures for modification, or other scientifically defensible methods. The USEPA develops aquatic life criteria, sediment guidelines for protection of benthic organisms, biological criteria, human health criteria for drinking water and ingestion of fish and shellfish, and WC. In the U.S. water program, WQC are used to evaluate whether a water body meets its designated use and to calculate discharge permit limits. Protective or remedial action may be needed if the WQC is exceeded.

Derivation Methodology

Aquatic life

The USEPA's aquatic life criteria consist of 3 components: magnitude, duration, and frequency. The magnitude of allowable pollutant concentration is determined from an examination of laboratory toxicity data for a variety of aquatic species.

Aquatic life criteria are derived using a minimum database of acute toxicity tests for at least 8 representative genera and at least 3 matched chronic toxicity tests. This database is used to represent the diversity and range of sensitivities of aquatic life. These data are extrapolated to a level intended to protect 95% of species from either acute or chronic adverse effects. Figure A-7 depicts the calculation procedure:

- Collect toxicity data for a minimum of 1 species in each of 8 different families (using the most sensitive life stage for each).
- Calculate the species mean acute value (SMAV) (mean of all acute results for each species).
- Calculate genus mean acute value (GMAV) (mean of all SMAVs within a genus).
- Rank the GMAVs from most to least sensitive.
- Use the lowest 4 GMAVs to calculate the final acute value (FAV).
- Extrapolate the 5th percentile from the 4 lowest GMAVs = FAV.
- Calculate the CMC - FAV / 2 to convert an LC50 to an LC1.

Chronic criteria are calculated similarly if data are available; if not, then an ACR is calculated from acute and chronic toxicity data. The FAV is divided by the ACR to derive the final chronic value (FCV). The criterion continuous concentration (CCC)

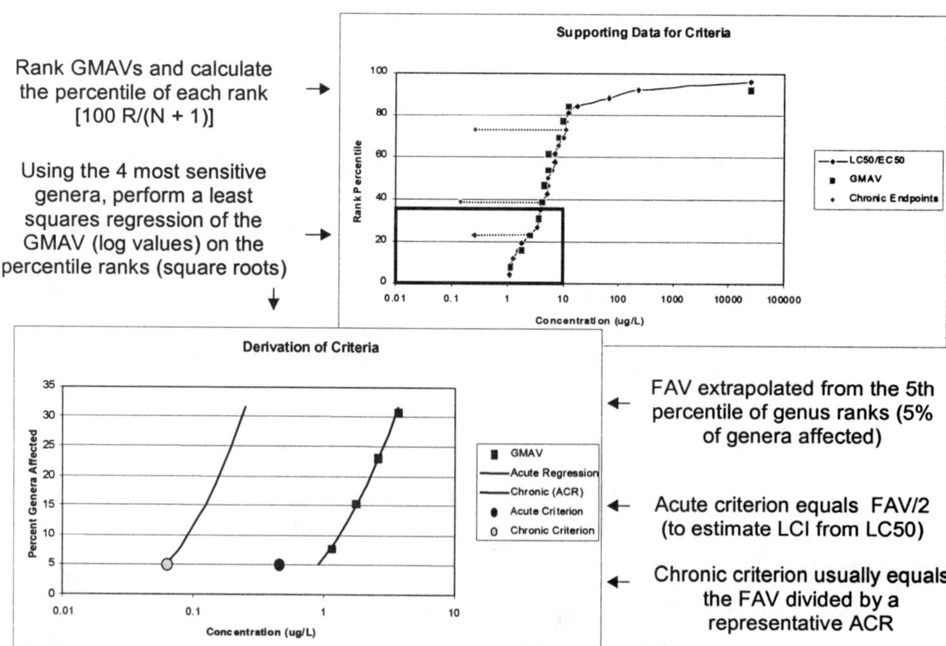

Figure A-7 Aquatic life WQC calculation example

is the lowest of the FCV, final plant value (FPV), final residue value (FRV), or maximum permissible tissue concentrations.

Allowances for duration of exposure and frequency of exceedance are also specified:
- CCC = 4 days (the average of all concentration of a pollutant measured within 4 days would not exceed the CCC).
- CMC = 1 hour (the average of all concentrations of a pollutant measured within 1 hour should not exceed the CMC).
- One excursion of both magnitude and duration in a 3-year period (assumes that systems have recovery potential if given the opportunity).

An example of an USEPA AWQC statement is as follows: The procedures described in the *Guidelines for Deriving Numerical National Water Quality for the Protection of Aquatic Organisms and Their Uses* indicate that, except possibly where a locally important species is very sensitive, freshwater aquatic organisms and their uses should not be affected unacceptably if the 4-day average concentration (in µg/L) of cadmium does not exceed the numerical value given by $e^{(0.7852[\ln(\text{hardness})] - 3.490)}$ more than once every 3 years on the average and if the 1-hour average concentration (in µg/L) of cadmium does not exceed the numerical value given by $e^{(1.128[\ln(\text{hardness})] - 3.828)}$ more than once every 3 years on the average. For example, at hardness of 50, 100, and 200 mg/L as $CaCO_3$ the 4-day average concentrations of cadmium are 0.66, 1.1, and 2.0 µg/L respectively, and the 1-hour average concentrations are 1.8, 3.9, and 8.6 µg/L.

Sediment guidelines are developed using the same intended level of protection used in WQC by employing equilibrium partitioning. USEPA has not finalized these values as of this writing, but previous draft publications indicate their derivation. For nonionic organic pollutants, the equation is:

$$\text{SQC} = \text{FCV (aquatic life)} \times \%\text{OC (of sediment)} \times K_{oc} \text{ (of chemical)} \qquad \text{(A-3)},$$

where
- FCV is the effects concentration of the chemical to aquatic life.
- %OC is the most significant binding component of the sediment; combined with K_{oc}, it accounts for the bioavailability of the chemical.
- K_{oc} establishes the distribution of the chemical between the OC and the interstitial water.

For metals mixtures (lead, nickel, cadmium, copper, zinc, silver), the equation is

$$\text{SEM} \leq \text{AVS } or \text{ [IW}_m] / \text{FCV} \leq 1 \qquad \text{(A-4)},$$

where

> SEM = metals released in a cold acid extraction; and
>
> AVS = (sulfide released during same extraction) the significant binding component in sediments for metals, accounts for the bioavailability of metals to benthic aquatic organisms.

The USEPA also intends to develop a sediment guideline for PAH mixtures that reflects the awareness that PAHs are rarely, if ever, found in the environment as individual contaminants and that their toxicities are additive. The PAH mixture guideline is based on the narcosis theory of toxic action and equilibrium partitioning with organic carbon.

Biological integrity

Biological criteria describe the biological integrity of aquatic communities. "Biological integrity" is defined as the condition of an aquatic community inhabiting unimpaired water bodies of a specified habitat, as measured by community structure and function. The development of biological criteria is based on the premise that the structure and function of an aquatic biological community within a specific type of water body provide critical information about the quality of surface waters. Biological criteria are developed from the assessment of unimpaired or minimally impaired water bodies (reference sites). Water bodies under evaluation are assessed and compared to the criteria to determine impairment and nonattainment of a designated use. The cause of the impairment has to be determined and remedies outlined and implemented using a variety of other tools.

Human health and wildlife

Criteria for consumers (humans and wildlife) are derived by determining protective tissue residue levels in food and using a BCF or, more recently, a BAF to relate to a water concentration. The complicating factor of bioaccumulation has only been taken into consideration in the development of criteria in the last several years. Both BCFs and BAFs have been developed for some chemicals, either on a site-specific empirical basis or calculated or modeled from available data. The BCF is a measure of a chemical's potential to bioaccumulate in the tissue of aquatic organisms and is a function of the lipid solubility of the chemical. The equation below can be used to empirically calculate the BCF, or it can be calculated from the log K_{ow}.

$$\text{BCF} = \frac{\text{Concentration of the chemical in tissue}}{\text{Concentration of chemical in water}} \tag{A-5}.$$

Because the BCF accounts only for the uptake of chemical from water and not from food, a BAF is also needed:

$$\text{BAF} = \text{FM} \times \text{BCF} \tag{A-6}.$$

The FM accounts for both the level of the food chain (1 for lower-order consumers [shellfish] and 4 for higher-order consumers [sport fish]) at which an organism feeds and the potential for a chemical to bioconcentrate in that organism.

The equation for a human health non-carcinogen criterion derivation is

$$\frac{(RfD \times WT) - (DT + IN) \times WT}{WI + [FC \times L \times FM \times BCF]} \qquad (A\text{-}7),$$

where

- RfD (reference dose) is the estimated daily exposure to a human population that is not likely to cause deleterious effects over a lifetime (milligrams of toxicant per kilogram of human body weight per day).
- WT is the weight of average human adult (70 kilograms).
- DT is the dietary exposure through food other than fish (milligrams of toxicant per kilogram of body weight per day).
- IN is the inhalation exposure (milligrams of toxicant per kilogram of human body weight per day).
- WI is the water intake of an average adult (2 liters).
- FC is the daily fish consumption (kilograms fish/day).
- L is the ratio of lipid fraction of fish tissue consumed to 3%.
- BCF is milligrams of toxicant per kilogram of fish, divided by milligrams of toxicant per liter of water.

Independent application: Putting it all together

In any given water body evaluation, the 3 general types of criteria—chemical, biological, and physical—must be independently evaluated. The failure of the water body to achieve any one of the three indicates the potential of impairment and non-attainment of uses.

The policy is held because each of the criteria types evaluates different concerns:
- chemical-specific criteria identify toxicity due to specific chemicals of concern;
- biological criteria (including whole effluent and whole sediment toxicity tests) identify impacts due to the integration of many environmental factors, including habitat, toxicity, water quality, and flow regime; and
- physical criteria identify impacts due to specific physical characteristic of the water (e.g., pH, hardness, solids, hydrology).

References

Aldenberg T. 1993. ETX 1.3a. A program to calculate confidence limits for hazardous concentrations based on small samples of toxicity data. Bilthoven, NL: National Institute of Public Health & Environment Protection. Nr 719102015.

Aldenberg T, Slob W. 1993. Confidence limits for hazardous concentrations based on logistically distributed NOEC toxicity data. *Ecotoxicol Environ Saf* 25:48-63.

[ANZECC] Australian and New Zealand Environment and Conservation Council. 1992. Australian water quality guidelines for fresh and marine waters. Canberra, Australia: ANZECC.

[ANZECC & ARMCANZ] Australian and New Zealand Environment and Conservation Council & Agriculture and Resource Management Council of Australia and New Zealand. 2000. Australian and New Zealand guidelines for fresh and marine water quality. Canberra, Australia: ANZECC & ARMCANZ. National Water Quality Management Strategy Paper nr 4.

[CCME] Canadian Council of Ministers of the Environment. 1991. Appendix IX—A protocol for the derivation of water quality guidelines for the protection of aquatic life (April 1991). In: [CCME] Canadian Council of Resource and Environment Ministers. 1987. Canadian water quality guidelines. Winnipeg MB, Canada: Task Force on Water Quality Guidelines. [Reprinted in Canadian environmental quality guidelines, Chapter 4, Canadian Council of Ministers of the Environment, 1999, Winnipeg.]

[CCME] Canadian Council of Ministers of the Environment. 1993. Appendix XV—Protocols for the derivation of water quality guidelines for the protection of agricultural water uses (October 1993). In: [CCME] Canadian Council of Resource and Environment Ministers. 1987. Canadian water quality guidelines. Winnipeg MB, Canada: Task Force on Water Quality Guidelines. [Reprinted in Canadian environmental quality guidelines, Chapter 5, Canadian Council of Ministers of the Environment, 1999, Winnipeg.]

[CCME] Canadian Council of Ministers of the Environment. 1995. Protocol for the derivation of Canadian sediment quality guidelines for the protection of aquatic life. Ottawa ON, Canada: Environment Canada, Guidelines Division, Technical Secretariat of the CCME Task Group on Water Quality Guidelines. CCME EPC-98E. [Reprinted in Canadian environmental quality guidelines, Chapter 6, Canadian Council of Ministers of the Environment, 1999, Winnipeg.]

[CCME] Canadian Council of Ministers of the Environment. 1996. A protocol for the derivation of environmental and human health soil quality guidelines. Winnipeg MB, Canada: CCME, [A summary of the protocol appears in Canadian environmental quality guidelines, Chapter 7, Canadian Council of Ministers of the Environment, 1999, Winnipeg.]

[CCME] Canadian Council of Ministers of the Environment. 1998. Protocol for the derivation of Canadian tissue residue guidelines for the protection of wildlife that consume aquatic biota. Winnipeg MB, Canada: CCME Water Quality Guidelines Task Group. [Reprinted in Canadian environmental quality guidelines, Chapter 8, Canadian Council of Ministers of the Environment, 1999, Winnipeg.]

[CCREM] Canadian Council of Resource and Environment Ministers. 1987. Canadian Water Quality Guidelines. Ottawa ON, Canada: Task Force on Water Quality Guidelines of the Council of Resource and Environment Ministers, Environment Canada.

[CCREM] Canadian Council of Resource and Environment Ministers. 1987. Canadian water quality guidelines. Winnipeg MB, Canada: Task Force on Water Quality Guidelines of the Canadian Council of Resource and Environment Ministers.

Crommentuijn T, Van de Plassche EJ, Canton JH. 1994. Guidance document on the derivation of ecotoxicological serious soil contamination in view of the intervention value for soil clean-up. Bilthoven NL: RIVM. Report nr 95001 003.

[CSR] Centre for Substances and Risk Assessments. 1996. QA-procedures for deriving environmental quality objectives (INS and I-values). CSR-KD/003.

Di Toro DM, Zarba CS, Hansen DJ, Berry WJ, Swartz RC, Cowman CE, Pavlou SP, Allen HE, Thomas NA, Paquin PR. 1991. Technical basis for establishing sediment quality criteria for nonionic organic chemicals using equilibrium partitioning. *Environ Toxicol Chem* 10:1541-1583.

Emans HJB, Van de Plassche EJ, Canton JH, Okkerman PC, Sparenburg PM. 1993. Validation of some extrapolation methods used for effect assessment. *Environ Toxicol Chem* 12:2139-2154.

Everts JW, Ruys M, Van de Plassche EJ, Pijnenburg J, Luttik R, Lahr J, Van der Valk, Canton JH. 1992. Doorvergiftiging in de voedselketen. Een route naar een maximaal toelaatbare concentratie in het mariene milieu. Dienst Getijdewateren and RIVM (in Dutch).

Health Canada. 1989. Guidelines for Canadian drinking water quality: Supporting documentation. Updated November 1990, December 1992, and February 1995. Ottawa. ON, Canada: Federal-Provincial Subcommittee on Drinking Water of the Federal-Provincial Committee on Environmental and Occupational Health.

Hickey CW, Pyle E. 2001. Derivation of water quality guideline values for heavy metals using a risk-based methodology: An approach for New Zealand. *Aust J Ecotoxicol* 7:137-156.

Janssen RPT, Peijnenburg WJGM, Posthuma L, van den Hoop MAGT. 1997. Equilibrium partitioning of heavy metals in Dutch field soils. I. Relationships between metal partition coefficients and soil characteristics. *Environ Toxicol Chem* 16: 2470-2478.

Janssen RPT, Posthuma L, Baerselman R, Den Hollander HA, Van Veen RPM, Peijnenburg WJGM. 1997. Equilibrium partitioning of heavy metals in Dutch field soils. II. Prediction of metal accumulation in earthworms. *Environ Toxicol Chem* 16:2479-2488.

Jongbloed RH, Pijnenburg J, Mensink BJWG, Traas ThP, Luttik R. 1994. A model for environmental risk assessment and standard setting based on biomagnification. Top predators in terrestrial ecosystems. Bilthoven, NL: RIVM. Report nr 719101012.

[MfE] Ministry for the Environment. 1996. A proposed methodology for deriving aquatic guideline values for toxic contaminants. Wellington, NZ: MFE. 56 p.

Pavlou SP, Weston DP. 1984. Initial evaluation of alternatives for development of sediment related criteria for toxic contaminants in marine waters (Phase II). Washington DC, USA: USEPA.

Peijnenburg WJGM, Posthuma L, Eijsackers HJP, Allen HE. 1997. A conceptual framework for implementation of bioavailability for environmental purposes. *Ecotoxicol Environ Saf* 37:163-172.

Posthuma L. 1997. Gradient studies as a possibility to associate ecotoxicological risk assessment with toxic effects on population and community parameters in the field. In: Løkke H, van Straalen NM, editors. Ecological principles for risk assessment of contaminants in soil. London, UK: Chapman and Hall.

Rainbow PS. 1993. The significance of trace metal concentrations in marine invertebrates. In: Dallinger R, Rainbow PS, editors. Ecotoxicology of metals in invertebrates. Boca Raton FL, USA: Lewis Publishers. p 3-23.

Romijn CAFM, Luttik R, Slooff W, Canton JH. 1992. Presentation of a general algorithm for effect assessment on secondary poisoning. Part 1, Aquatic foodchains. Bilthoven, NL: RIVM Report nr. 679102006.

Scheinberg H. 1991. Copper. In: Merian E. editor. Metals and their compounds in the environment: Occurrence, analysis and biological relevance. Weinheim, Germany: Wiley-VCH. p 893-908.

Shea D. 1988. Developing national sediment quality criteria. *Environ Sci Technol* 22:256-1261.

Slooff W. 1992. RIVM Guidance document. Ecotoxicological effect assessment: Deriving maximum tolerable concentrations (MTC) from single-species toxicity data. Bilthoven, NL: RIVM Report nr.719102 018.

Struijs J, van de Meent D, Peijnenburg WJGM, Crommentuijn T. 1997. Added risk approach to derive maximum permissible concentrations for heavy metals: How to take into account the natural background levels. *Ecotoxicol Environ Saf* 37:112-118.

Van de Meent D, Aldenberg T, Canton JH, Van Gestel CAM, Slooff W. 1990. Desire for levels. Back-ground study for the policy document, Setting environmental quality standards for water and soil. Bilthoven, NL: RIVM Report nr. 670101 002.

Van de Meent D, De Bruijn JHM. 1995. A modeling procedure to evaluate the coherence of independently derived environmental quality objectives for air, water and soil. *Environ Toxicol Chem* 14:177-186.

Van de Plassche EJ. 1994. Towards integrated environmental quality objectives for several compounds with a potential for secondary poisoning. Bilthoven, NL: RIVM Report nr. 679101 012.

[VROM] Dutch Ministry of Housing, Physical Planning and Environment. 1989. Premises for risk management. Risk limits in the context of environmental policy. Second chamber, session 1988-1989, 21137, no 5.

[VROM] Dutch Ministry of Housing, Physical Planning and Environment. 1994. Environmental quality objectives in the Netherlands. The Hague, NL: VROM.

Warne M StJ. 1998. Critical review of methods to derive water quality guidelines for toxicants and a proposal for a new framework. Canberra, Australia: Office of the Supervising Scientist. Report 135.

[WQAQOG] Federal-Provincial Working Group on Air Quality Objectives and Guidelines. 1996. A protocol for the development of national ambient air objectives, Part 1, Science assessment document and derivation of the reference level(s). Toronto and Ottawa, ON, Canada: Environment Canada and Health Canada. Catalogue #En42-17/5-1-1997E.

A Risk Assessor's Thoughts on Water-Quality Criteria Development

John E. Toll

In a study to help reevaluate the WQC for selenium, we used tissue residue effects data and ERA methods (Adams et al. 1998). The conclusions of that study led us to develop a risk-based methodology for developing site-specific selenium criteria using tissue residue data. Following is a brief description of the work. I would like to begin by acknowledging my colleagues who have been involved in one or more aspects of this work: Bill Adams, Kennecott Utah Copper, Inc.; Rick Bennett, ecological planning and toxicology, inc.; Kevin Brix, Rick Cardwell, and Kristin Cothern, Parametrix Inc.; Maxine Dakins, University of Idaho; David DeForest, Parametrix Inc.; Anne Fairbrother, ecological planning and toxicology, inc.; Mitchell Small, Carnegie Mellon University; and Lucinda Tear, Parametrix Inc.

Selenium poisoning was observed in Belews Lake, North Carolina, USA, in the mid 1970s and in Kesterson Reservoir, California, USA, in the mid 1980s. These observations led to numerous studies evaluating the potential for selenium to cause ecologically significant effects via food-chain transfer in aquatic ecosystems, especially wetlands. The current national chronic ambient WQC for protection of aquatic life is 5 mg/L. Scientists with the U.S. Fish and Wildlife Service (USFWS) have recommended setting the ambient WQC at 2 mg/L for both aquatic and wildlife protection.

Reported site-specific variations in selenium's effects on aquatic life and birds prompted us to reevaluate the basis for the 2 mg/L recommendation (Adams et al. 1998). We used probabilistic regression models to assess water, food-chain, and bird egg residues from 15 lentic sites in the western U.S. Uncertainty analysis of the regression models provided a probability distribution of waterborne selenium concentrations associated with bird egg tissue residues. Using the 10th and 50th percentiles of these distributions, we calculated waterborne selenium concentrations between 6.8 and 46 mg/L that are protective of birds at different sites. These values are associated with an effect threshold of 20 mg/kg selenium dry weight in bird eggs, which is the EC10 for mallard duck embryo teratogenesis (Skorupa 1996). The water concentrations protective of birds range from slightly more than the current USEPA WQC (6.8 versus 5.0 mg/L) to a factor of 10 or greater at some sites.

Overall, Adams et al. (1998) found a high correlation between water and mean egg selenium concentrations that is strongly influenced by site-specific factors. Key factors influencing potential for site risk are selenium speciation, redox conditions, and extent of biological receptor co-occurrence. In light of the observed site variability, we concluded the following. First, the numerical WQC for selenium should be used only as a screening tool. Second, when waterborne selenium approaches or exceeds the criterion, a site-specific assessment should be used to determine whether existing water concentrations pose risk, and if so to identify a safe selenium water concentration for the site. We developed a methodology for doing the type of site-specific assessment we envisioned when we reached these conclusions. It uses data from a specific site to calibrate a model developed with data from other similar sites. The logic behind the methodology is as follows:

When one looks at a specific site, one typically observes a limited range of water selenium concentrations—sometimes only a single concentration[1]—and a limited number of tissue (e.g., egg) samples.

1) Because of the limited range of water concentrations, and the uncertainty in the mean tissue residue concentration, it is difficult to extrapolate these data to a water concentration that results in an average tissue residue concentration less than or equal to an effect threshold.

2) Data from other similar sites can provide a broader range of water and tissue residue concentrations, but we know there is significant site-to-site variability in the relationship between water and mean tissue residue.

3) We can use the pooled data from other similar sites to define a set of possible water and mean tissue residue relationships, then use our site-specific data to determine which relationships, from the set of possibilities, fit our specific site.

4) Once we have determined which of the set of possible relationships fits our specific site, we can extrapolate from the observed water concentration to a water concentration that results in a tissue residue concentration less than or equal to an effect threshold. The result is a site-specific, tissue residue-based WQC.

---

[1] When one is comparing a water concentration to a tissue residue concentration, the tissue residue represents integrated exposure, so it should be compared to a temporally and spatially averaged water concentration. This is why one might expect to have only a single water concentration at a specific site.

A useful property of the methodology we have developed is that the better the site-specific data, the more able it is to reduce the range of relationships possibly fitting the specific site, and the smaller the margin of safety it places on the site-specific criterion.

The statistical technique we use in our methodology is Bayesian Monte Carlo (BMC) analysis. Monte Carlo methods are numerical techniques for generating a representative sample from a probability distribution function (PDF). BMC evolved from earlier procedures used to ensure that the PDF of a model's predictions was consistent with observed data. These earlier acceptance or rejection procedures involved deleting predictions that were inconsistent with observations (Hornberger and Spear 1980; Beck 1987; Woodruff et al. 1992). The difference between BMC and earlier procedures is that BMC is more statistically rigorous. BMC defines the acceptance or rejection procedure using the axioms of probability theory, as expressed in Bayes' theorem.[2]

The use of Bayes' theorem to define the acceptance or rejection procedure for evaluating a Monte Carlo PDF was first illustrated by Dilks et al. (1989, 1992), who coined the term "BMC." Other applications include Patwardhan and Small (1992), Small and Escobar (1992), and Dakins et al. (1996). Other investigators were using Bayesian techniques in the late 1980s and early 1990s (e.g., Erdy 1989; Iman and Hora 1989; Wolpert et al. 1992), but the techniques were computationally limited until combined with Monte Carlo analysis.

Adams et al. (1998), in their regression model of lentic sites in the western U.S., demonstrated that whenever one is working with a selenium bioaccumulation model that is not site specific, even if water selenium concentration is constant, the mean egg selenium prediction should be expected to be probabilistic because of site-to-site variability in the water–mean egg selenium relationship. For example, suppose a water selenium concentration of 2 mg/L is entered into the model. At this water concentration, the model for a particular bird species might predict a normally distributed mean egg selenium concentration with a mean of 4 mg/kg selenium dry weight in bird eggs, and a standard deviation of 0.4 mg/kg. This normal PDF would represent the site-to-site variability in the water to mean egg selenium relationship for that bird species, at lentic sites in the western U.S.

---

[2] Bayes' theorem states that the probability associated with a prediction, given some new data, is proportional to the probability of the prediction before the new data were available, times the probability of having observed the new data if the prediction were true. The proportionality constant in Bayes' theorem ensures that the sum of the probabilities of all possibilities is 1. Bayes' theorem is named after Rev. Thomas Bayes, an 18th century mathematician who derived a special case of this theorem (Bayes 1763). The theorem was generalized by Laplace (1814). It is the basic starting point for inference problems using probability theory as logic (Bretthorst 1994; Jaynes 1994). It is a simple derivation from the axioms of probability theory and can be found near the beginning of many probability and statistics textbooks.

The previous paragraph argues that a single water selenium concentration put into a nonsite-specific model should produce a PDF of mean egg selenium predictions. Generally speaking, observed mean egg selenium concentrations are uncertain, too. Observations are uncertain due to sampling and measurement error. For example, one may have a random sample of eggs collected from a specific site, with a sample average egg selenium concentration of 2 mg/kg and a sample standard error of 1 mg/kg.

Bayesian Monte Carlo analysis accounts for both the uncertainty in the mean egg selenium predictions (from site-to site variability) and the uncertainty in mean egg selenium observations (from sampling and measurement error). The BMC methodology asks, given the uncertainties in the site-specific data, which of the model's predictions could be representative of the site? In other words, it uses the data from multiple sites to define the range of possible water–mean egg selenium relationships, then uses data from a specific site to narrow down the set of relationships that could apply to that site. Some predictions of mean egg selenium concentration will be more similar to the observed mean egg selenium concentration at a specific site than others. BMC statistically compares the uncertain predictions of the nonsite-specific model to the site-specific data, and measures the degree of similarity between the predictions and observations. It updates the model by increasing the probabilities on predictions that are more consistent with observations, and reducing the probabilities on predictions that are less consistent with observations. In other words, it places more weight on predictions that are more consistent with site-specific observations and less weight on predictions that are less consistent with the observations. Therefore, BMC uses observed data to make predictive models more site specific.

We applied BMC to an assessment of the water–mean egg selenium relationship for a wetland on the south shore of the Great Salt Lake (USA). We used data from the southshore wetland and BMC to develop a site-specific, probabilistic prediction of mean egg selenium as a function of the average (by nest) water selenium concentration to which birds were exposed. This work was completed as part of a selenium ERA for the southshore wetland (Kennecott Utah Copper 1998). Earlier, we described our conclusions about using selenium criteria:

1) the numerical WQC for selenium should be used only as a screening tool, and

2) when waterborne selenium approaches or exceeds the criterion, a site-specific assessment should be used to determine a) whether existing water concentrations pose risk, and if so b) to identify a safe selenium water concentration for the site.

We did the ERA because waterborne selenium at the site exceeded the current national chronic AWQC of 5 mg/L. As we will describe, the site-specific assessment found minimal risk to birds from selenium at the site. Therefore, it was not necessary in this case to proceed to Step 2(b) and identify a safe selenium water concentration for the site.

The next 3 paragraphs briefly describe the exposure assessment, effect characterization, and risk characterization for the southshore wetland risk assessment.

The exposure analysis provided mean egg selenium PDFs for stilts and avocets. We developed these using all available information on shorebird selenium exposure. The data included measured macroinvertebrate and egg selenium concentrations, locations of nests on the southshore wetlands, as well as probabilistic selenium trophic transfer factors (TTFs), which are statistically derived ratios of mean egg to mean invertebrate selenium concentrations. We derived the TTF PDFs using data compiled by Adams et al. (1998) for the 15 lentic sites in the western U.S. We multiplied these by the PDF of site-specific macroinvertebrate selenium concentrations to get initial PDFs of mean stilt and avocet egg selenium concentrations. We combined the initial PDFs with the site-specific mean egg selenium and nest location data using BMC. The result is the calibrated—or "posterior"—mean egg selenium PDF, which gives a statistical representation of variability and uncertainty in stilt and avocet exposure to water selenium at the southshore wetlands. The prior and calibrated cumulative PDFs for black-necked stilts are shown in Figure A-8. Variability in exposure arose from variability in water selenium concentrations across the southshore wetlands. Uncertainties were associated with the spatial pattern of habitat used by nesting shorebirds, their diets, and TTFs.

The effect characterization used egg teratogenesis as the measure of effect. We used the mean egg selenium–teratogenicity dose–response model of Ohlendorf et al. (1986). Their dose–response model predicts the percentage of eggs exhibiting teratogenesis as a function of mean egg selenium concentration. Teratogenicity was selected as the measure of effect because it was the most sensitive endpoint with sufficient data to derive a dose–response relationship.

The risk characterization ran the mean egg selenium PDFs through the dose–response model, predicting probability of teratogenic effects as a function of mean egg selenium concentration for stilts and avocets. Results from this risk assessment indicate 0.1% to 0.5% of avocet and 2.4% to 3.8% of stilt chicks were expected to be teratogenically affected by selenium in the breeding season for which the data were collected.[3] Considering the number of birds nesting in the southshore wetlands and the relative hatching success independent of selenium effects, only 0 to 1 avocet and 4 to 7 stilts chicks are expected to be teratogenically effected by selenium. This compared to the 300+ eggs per year for each species that naturally do not survive in the southshore wetlands due to factors such as nest flooding and predation. Therefore, we concluded that risks from selenium to aquatic birds nesting in the

---

[3] Since the data were collected, a groundwater source of selenium has been identified and controlled, so the conditions represented in the data differ from current conditions.

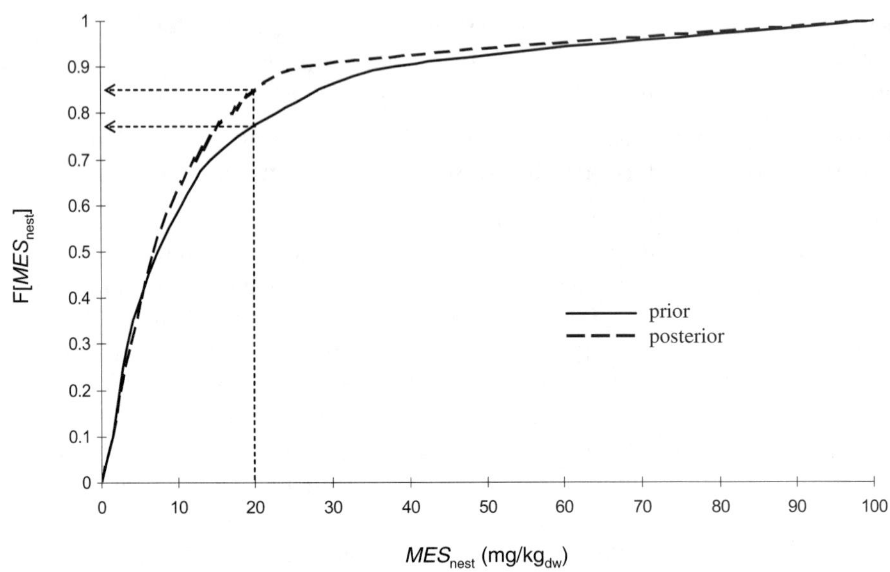

Figure A-8 The prior and posterior cumulative PDFs for nest-averaged egg selenium concentration in Kennecott Southshore Wetlands black-necked stilts. The dashed lines and arrows provide examples of how to read this figure. Before the model was calibrated with site-specific egg selenium and nest location data, 23% ([1 – 0.77] X 100%) of possible nests were estimated to have an average egg selenium concentration above 20 mg/kg_dw. After the model was calibrated, the estimate dropped to 14%. When combined with a dose–response model (Ohlendorf et al. 1986), the posterior distribution shown here corresponded to an estimate of 2 to 4% teratogenicity in black-necked stilt chicks, versus an estimated 8% teratogenicity predicted with the prior (uncalibrated) model. (MES = mean egg selenium concentration)

southshore wetlands were minimal. Considering the concentrations and geographical distribution of selenium in the wetland, very few chicks had the potential to be teratogenically affected.

The southshore wetland case study provided an interesting opportunity to demonstrate a risk-based methodology for developing site-specific selenium criteria using tissue residue data. We would also like to make some general observations about risk-based methods for setting site-specific criteria, in particular using tissue residue data. First, tissue residue-based WQC are not necessary for all stressors, but we need to consider them for bioaccumulative chemicals because protection of aquatic life and human health is less likely to protect ecosystems from the effects of these stressors. Second, as we move toward tissue residue-based WQC, we have to assume that site-specific factors will govern the relationship between chemical concentrations in waters or sediments and tissue residues. The importance of site-specific factors will necessitate a risk-based approach to developing tissue residue-based criteria. The risk-based approach provides a framework for setting site-specific management goals, identifying assessment endpoints, and conducting the site-specific assessment in a way that can foster the cost-effective development of WQC that are protective of valued populations, communities, and ecosystem functions.

References

Adams WJ, Brix KV, Cothern KA, Tear LM, Cardwell RD, Fairbrother A, Toll JE. 1998. Assessment of selenium food chain transfer and critical exposure factors for avian wildlife species: Need for site-specific data. In: Little EE, DeLonay AJ, Greenberg BM, editors. Environmental toxicology and risk assessment. 7th volume. Philadelphia PA, USA: ASTM. STP 1333.

Bayes RT. 1763. An essay toward solving a problem in the doctrine of chances. *Philos Trans R Soc Lond Biol Sci* 53:370-418. Reprinted in *Biometrika* 45:293-315 (1958).

Beck MB. 1987. Water quality modeling: A review of the analysis of uncertainty. *Water Resour Res* 23:1393-1442.

Bretthorst GL. 1994. An introduction to model selection using probability theory as logic. In: Heidbreder G, editor. Maximum entropy and Bayesian methods. Dordrecht, NL: Kluwer Academic Publishers.

Dakins ME, Toll JE, Small MJ, Brand K. 1996. Risk-based environmental remediation: Bayesian Monte Carlo analysis and the expected value of sample information. *Risk Anal* 16:67-69.

Dilks DW, Canale RP, Meier PG. 1989. Analysis of model uncertainty using Bayesian Monte Carlo. Proceedings of ASCE Specialty Conference on Environmental Engineering; American Society of Civil Engineers; 1989; New York NY, USA. p 571-577.

Dilks DW, Canale RP, Meier PG. 1992. Development of Bayesian Monte Carlo techniques for water quality model uncertainty. *Ecol Model* 62:149-162.

Erdy DM. 1989. The confidence profile method: A Bayesian method for assessing health technologies. *Oper Res* 37:210-228.

Hornberger GM, Spear RC. 1980. Eutrophication in Peel Inlet II: Identification of critical uncertainties via generalized sensitivity analysis. *Water Res* 14:43-49.

Iman RL, Hora SC. 1989. Bayesian methods for modeling recovery time with an application to the loss of off-site power at nuclear power plants. *Risk Anal* 9:25-36.

Jaynes ET. 1994. Probability theory: The logic of science. Available at http://omega.albany.edu:8008/JaynesBook.html. Accessed 14 Mar 2003.

Kennecott Utah Copper. 1998. Selenium risk assessment for Kennecott southshore wetlands. Prepared for Kennecott Utah Copper, Inc. by Parametrix, Inc. and ecological planning and toxicology, inc.

Laplace PS. 1814. A philosophical essay on probabilities. Unabridged and unaltered reprint of Truscott and emory translation. New York NY, USA: Dover Publications, Inc., 1951.

Ohlendorf HM, Hoffman DJ, Saiki MK, Aldrich TW. 1986. Embryonic mortality and abnormalities of aquatic birds: Apparent impacts of selenium from irrigation drainwater. *Sci Total Environ* 52:49-63.

Patwardhan A, Small MJ. 1992. Bayesian methods for model uncertainty analysis with applications to future sea level rise. *Risk Anal* 12:513-523.

Skorupa JP. 1996. Avian selenosis in nature: A phylogenetic surprise and its implications for ecological risk assessment. Presented at the 17th Annual Meeting of the SETAC; 1996 Nov 15–21; Washington DC, USA.

Small MJ, Escobar MD. 1992. Discussion of paper by Wolpert et al. In: Gastonis C, Hodges XJ, Kass R, Singpurwalla N, editors. Bayesian statistics and technology: Case studies. New York NY, USA: Springer-Verlag.

Wolpert RL, Steinberg LJ, Reckhow KH. 1992. Bayesian decision support using environmental transport and fate models. In: Gastonis C, Hodges XJ, Kass R, Singpurwalla N, editors. Bayesian statistics and technology: Case studies. New York NY, USA: Springer-Verlag.

Woodruff TJ, Bois FY, Auslander D, Spear RC. 1992. Structure and parameterization of pharmacokinetic models: Their impact on model prediction. *Risk Anal* 12:189-210.

Abbreviations

7Q10	7-day average low flow expected to occur once in 10 years
AChE	acetylcholinesterase
ACR	acute-to-chronic ratio
AF	application factor
AhR	Ah receptor
ANOVA	analysis of variance
ANZECC	Australian and New Zealand Environment and Conservation Council
AQUIRE	Aquatic Information Retrieval
ARMCANZ	Agriculture and Resource Management Council of Australia and New Zealand
ASFA	Aquatic Sciences and Fisheries Abstracts
ASTM	American Society for Testing and Materials
AVS	acid volatilized sulfide
AWQC	ambient water-quality criteria
BACI	before–after control–impact
BAF	bioaccumulation factor
BCF	bioconcentration factor
BLM	Biotic Ligand Model
BMC	Bayesian Monte Carlo
BSAF	Biota–sediment accumulation factor
CBR	critical body residue
CCC	criteria continuous concentration
CCME	Canadian Council of Ministers of the Environment
CCREM	Canadian Council of Resource and Environment Ministers
ChV	chronic value
CMC	criterion maximum concentration
CSO	combined sewer overflow
CWQG	Canadian Water Quality Guidelines
DDE	dichlorodiphenyl dichloroethylene
DDT	dichlorodiphenyl trichloroethane
DNR	Department of Natural Resources
DOC	dissolved organic carbon
DOM	dissolved organic matter
ECOTOX SCC	ecotoxicological serious contamination concentration
EP	equilibrium partition
EQG	environmental quality guideline
EQS	environmental quality standard
ER	estrogen receptor
ERA	ecological risk assessment
ERL	ecotoxicological risk limit
ESA	Endangered Species Act (U.S.)
FAV	final acute value
FCV	final chronic value

Reevaluation of the State-of-the Science for Water-Quality Criteria Development. Mary C. Reiley et al., editors.
©2003 Society of Environmental Toxicology and Chemistry (SETAC). ISBN 1-880611-30-9

FIFRA	Federal Insecticide, Fungicide, and Rodenticide Act (U.S.)
FM	food-chain multiplier
FPV	final plant value
FRV	final residue value
GMAV	genus mean acute value
HUMTOX SCC	human toxicological serious soil contamination concentration
K_{DOC}	partition coefficient for DOC
K_{oc}	organic carbon partition coefficient
K_{ow}	octanol–water partition coefficient
LC50	lethal concentration for 50% of a population
LDx	lethal dose for x% of a population
LOAEC	lowest-observed-adverse-effect concentration
LOEC	lowest-observed-effect concentration
LOEL	lowest-observed-effects level
MfE	Ministry for Environment (New Zealand)
MOA	mode of action
MPC	maximum permissible concentration
MSD	minimum significant difference
NC	negligible concentration
NOAEC	no-observed-adverse-effect concentration
NOEC	no-observed-effect concentration
NOEL	no-observed-effect level
NPDES	National Pollutant Discharge Elimination System (U.S.)
OECD	Organization for Economic Cooperation and Development
OM	organic matter
PAH	polycyclic aromatic hydrocarbon
PB-TK	physiologically based toxicokinetic
PCB	polychlorinated biphenyl
PCDD	polychlorinated dibenzo-p-dioxin
PCDF	polychlorinated dibenzofuran
PDF	probability distribution function
POC	particulate organic carbon
QSAR	quantitative structure–activity relationship
RMA	Resource Management Act (New Zealand)
SAR	structure–activity relationship
SEM	simultaneously extracted metal
SETAC	Society of Environmental Toxicology and Chemistry
SMAV	species mean acute value
SQC	sediment-quality criterion or criteria
SQG	sediment-quality guideline
TCDD	tetrachlorodibenzo-p-dioxin
TEC	toxic equivalent concentration
TEF	toxic equivalency factor
TMDL	total maximum daily load
TSCA	Toxic Substances Control Act (U.S.)

TTF	trophic transfer factor
USEPA	U.S. Environmental Protection Agency
USFWS	U.S. Fish and Wildlife Service
WC	wildlife criteria
WER	water–effect ratio
WET	whole effluent toxicity
WGAQOG	Working Group on Air Quality Objectives and Guideline(Canada)
WQC	water-quality criterion or criteria
WQG	water-quality guideline
WQS	water-quality standard

Index

Burial velocities, 43

C

Canadian Council of Ministers of the Environment (CCME), 144

Canadian water-quality guidelines (CWQGs), 144
    derivation protocol, 146
    guideline development protocol for, 148

Carrier proteins, 21

Catastrophic stresses, rates of recovery from, 30

Cationic metals, 74, 89

CBRs. *See* Critical body residues

CCC. *See* Criterion continuous concentration

CCME. *See* Canadian Council of Ministers of the Environment

Ceriodaphnia sp. reproduction test, 98

Chemical(s)
    action of by narcosis, 56
    bioaccumulation characteristics of, 34
    bioavailability, 6, 73
    biological response to, 69
    classes
       according to mode of toxic action and environmental behavior, 4
       criteria, 84
       environmental characteristics of, 9
    discharge, biological effects of, 41
    effects, structural measures of, 102
    field testing of, 123
    freely dissolved, 80
    loading, effects of cumulative, 122
    monitoring for newly developed, 46
    partitioning, variations in, 15
    phytotoxic, sensitivity of algal species to, 87
    residues, in eggs, 39
    single-medium criterion for, 44
    slowly accumulating, 72
    specificity, expression of, 44
    stress, recovery of aquatic ecosystems from, 105, 106
    stressors, 123
    waterborne, factors modifying bioavailability, 73
    water column, 122

Chromium
    oxidation state for, 18
    summary data for in freshwater, 158

Chronic criteria, lowering of, 65

Chronic data set(s)
    characteristics of, 157
    limitations, correction of, 65

Chronic effect thresholds, definition of, 69

Chronic endpoints, ecological significance of, 66

Chronic exposure–response relationships, advantages, 62

Chronic toxicity
    data, availability of for range of organisms, 65
    endpoints, 60
    quantification of, 60
    tests, 25
    time-dependence of, 72

Chrysemys picta, 95

Cladocerans
    mortality of, 129
    sensitivity of to organophosphate pesticides, 57, 129

Clean Water Act, 162

CMC. *See* Criteria maximum concentrations

Combined sewer overflow (CSO), 124

Community(ies)
    composition, shift in, 102
    -level risks, 12
    resiliency, 128
    structure and function of across ecosystems, 5
    water supplies, 145

Compounds, tendency of to biomagnify, 38

Computer models, 4

Concentration addition models, 90

Conceptual model(s)
    development of to describe additive toxicity, 89–90
    difference between Type 1 and Type 2, 10
    types of criteria, 6
    use of to derive criteria, 3

Confounding stressors, 126

Consumers, criteria for, 166

Contaminant(s)
    assessment of toxic effects of, 153
    dissolved, 43
    effects of on aquatic organisms, 101
    exposure, principal route of, 1
    fate models, 45
    inter-media transfer of, 41
    loss, by degradation, 43
    maternal transfer of, 35

Lowest-observed-effects level (LOEL), 146

M

Macroalgae, phylogenetic and morphologic
diversity among, 87
Macroinvertebrates, recovery potential of, 131
Macrophytes, 128
Mallard(s), 95
    duck embryo teratogenesis, 172
    LD50 protocols for, 94
Mammals
    data on sensitivity of, 151
    determinants of chemical accumulation
      by, 33
    exposure of to bioaccumulative chemicals,
      32
Management goal, definition of, 124
Margins of safety, 147
Marine organisms, guideline values calculated
    for, 161
Marine taxa, laboratory testing of, 85
Mass–balance equation, sediment, 43
Maximum permissible concentrations (MPCs),
    7
Measurement endpoints, 5
Mechanistic models
    development of to replace BAF concept,
      39
    use of to predict bioavailability, 17
Media
    guidelines for, 148
    relationships among different, 10, 44
    -specific criteria, 17
Mercury uptake, 38
Mesocosm(s), 101, 104
    exposure–response experiments with, 129
    studies, endpoints examined in, 126
Metal(s)
    aquatic life effects for, 20
    assimilation, general principles governing,
      33
    binding
      kinetics, 76
      proteins, 34
    bioavailability, 22, 31
    BLM incorporated into WQC for, 23
    cationic, 74, 89
    criteria, development of, 18
    dietary, 33, 98
    dissolved, 74

    factors influencing chemical form of in
      freshwater, 19
    guideline values calculated for selection
      of, 160
    incorporation of BLM into WQC for, 45
    –receptor interaction, 22
    selection of priority, 154
    toxicity, 11
    transition, chemical speciation of, 23
    uptake model, 76
Metalloids, 78, 79
Metallothionein, 34, 93
Metapopulation models, 130
Methylation, requirement of in model to
    predict methyl mercury accumulation in
    fish, 23
Methylmercury, 11, 96
Microalgae, phylogenetic and morphologic
    diversity among, 87
Microcosm(s), 101, 104
    diazinon, 130
    experiments, 106
    exposure–response experiments with, 129
Micronutrients, taking up of by aquatic
    organisms, 75
Migration rates, 125
Migratory species, residues contained in, 93
Minerals, common constituents of, 78
Minimum significant difference (MSD), 62
Mining, 79
Mink
    crayfish preyed upon by, 92
    LD50 protocols for, 94
    reproductive effects of PCBs in, 95
    reproductive failure in, 91
    tests, protocols for acute and chronic, 95
Mixing zones
    acute, 27
    regulations of, 27
MOA. *See* Mode of action
Mode of action (MOA), 3
    chemicals operating through similar, 90
    presumed, 59
Model(s)
    accumulation, kinetic-based, 17
    acute toxicity versus time, 28
    bioaccumulation
      dietary exposure and, 17
      multipathway, 42
    computer, 4

O

Octanol–water partition coefficient (K_{ow}), 20, 37
OECD. *See* Organization for Economic Cooperation and Development
OM. *See* Organic matter
Opsanus tau, 98
Organic matter (OM), 19, 21
Organic partitioning, 31
Organization for Economic Cooperation and Development (OECD), 94, 154, 155
Organochlorine compounds, acute lethality of birds and brain residues of, 38
Organometallics, 78
Organophosphate pesticides, sensitivity of crustaceans to, 57, 129
Organoselenium, uptake of by waterfowl and fish, 18
Oxidation state, 18,

P

PAHs. *See* Polycyclic aromatic hydrocarbons
Painted turtle, 95
Pair feeding, experimental designs incorporating, 99
Paracellular pathways, leaky, 75
Parasites, susceptibility to, 68
Particulate organic carbon (POC), 79, 80
Passerines, forms of aquatic life as food source for, 91
Pathogens, susceptibility to, 68
PB-TK models. *See* Physiologically based toxicokinetic models
PCBs. *See* Polychlorinated biphenyls
PCDDs. *See* Polychlorinated dibenzo-*p*-dioxins
PCDFs. *See* Polychlorinated dibenzofurans
PDF. *See* Probability distribution function
Peer review, 121
Pentachlorophenol, ionized form of, 81
Performance indicators, Type 3 criteria used as, 124
Periphyton, 128
Permit
    limits, calculating of, 5
    writing, 8
Peromyscus maniculatus, 95
Pesticide(s)
    insect growth regulators developed as, 57–58

organophosphate, sensitivity of crustaceans to, 57
    registration, guidelines for, 87
    sensitivity of Glochidia to, 85
    testing, 125
pH, effects of on bioavailability, 9
Pharmacokinetic models, 70
Phenols, speciation of, 20
Physiologically based toxicokinetic (PB-TK) models, 41
Phytoplankton, 128
    dynamics, changes in, 86
    generation times, 29
    recovery potential of, 131
    sampling of natural assemblages of, 86
    toxicity of copper to, 19
Phytotoxic chemicals, sensitivity of algal species to, 87
Phytotoxicity, 86
Piscivorous birds
    exposure of to bioaccumulative chemicals, 32
    risk to, 11
Plankton, lake trout fry feeding on, 35
Plants
    survival and growth tests with, 59
    toxicity of boron to, 79
POC. *See* Particulate organic carbon
Point source
    controls, effectiveness of, 147
    discharge, hydrodynamic changes due to modifying of, 12
Pollutant(s)
    biological responses to, 17
    non-bioaccumulative, 30
    transformation of in environment, 24
Pollution indices, 102
Polychlorinated biphenyls (PCBs), 45, 57, 89, 96
    birds at risk for, 11
    consumption of fish containing, 91
    reproductive effects of in mink, 95
Polychlorinated dibenzofurans (PCDFs), 57
Polychlorinated dibenzo-*p*-dioxins (PCDDs), 57
Polycyclic aromatic hydrocarbons (PAHs), 89
    joint toxicity of, 91
    mixtures, sediment guidelines for, 166
Population(s)
    benthic invertebrate, 127

SETAC

A Professional Society for Environmental Scientists and Engineers and Related Disciplines
Concerned with Environmental Quality

The Society of Environmental Toxicology and Chemistry (SETAC), with offices currently in North America and Europe, is a nonprofit, professional society established to provide a forum for individuals and institutions engaged in the study of environmental problems, management and regulation of natural resources, education, research and development, and manufacturing and distribution.

Specific goals of the society are:
- Promote research, education, and training in the environmental sciences.
- Promote the systematic application of all relevant scientific disciplines to the evaluation of chemical hazards.
- Participate in the scientific interpretation of issues concerned with hazard assessment and risk analysis.
- Support the development of ecologically acceptable practices and principles.
- Provide a forum (meetings and publications) for communication among professionals in government, business, academia, and other segments of society involved in the use, protection, and management of our environment.

These goals are pursued through the conduct of numerous activities, which include:
- Hold annual meetings with study and workshop sessions, platform and poster papers, and achievement and merit awards.
- Sponsor a monthly scientific journal, a newsletter, and special technical publications.
- Provide funds for education and training through the SETAC Scholarship/Fellowship Program.
- Organize and sponsor chapters to provide a forum for the presentation of scientific data and for the interchange and study of information about local concerns.
- Provide advice and counsel to technical and nontechnical persons through a number of standing and ad hoc committees.

SETAC membership currently is composed of more than 5,000 individuals from government, academia, business, and public-interest groups with technical backgrounds in chemistry, toxicology, biology, ecology, atmospheric sciences, health sciences, earth sciences, and engineering.

If you have training in these or related disciplines and are engaged in the study, use, or management of environmental resources, SETAC can fulfill your professional affiliation needs.

All members receive a newsletter highlighting environmental topics and SETAC activities, and reduced fees for the Annual Meeting and SETAC special publications.

All members except Students and Senior Active Members receive monthly issues of *Environmental Toxicology and Chemistry* (ET&C), a peer-reviewed journal of the Society. Student and Senior Active Members may subscribe to the journal. Members may hold office and, with the Emeritus Members, constitute the voting membership.

If you desire further information, contact the appropriate SETAC Office.

SETAC North America
1010 North 12th Avenue
Pensacola, Florida 32501-3367 USA
T 850 469 1500   F 850 469 9778
E setac@setac.org

SETAC Europe
Avenue de la Toison d'Or 67
B-1060 Brussels, Belgium
T 32 2 772 72 81   F 32 2 770 53 83
E setac@setaceu.org

www.setac.org

Environmental Quality Through Science®